Keio University Symposia
for Life Science and Medicine 5

Springer

Tokyo
Berlin
Heidelberg
New York
Barcelona
Hong Kong
London
Milan
Paris
Singapore

Y. Ikeda, J. Hata, S. Koyasu
Y. Kawakami, Y. Hattori (Eds.)

Cell Therapy

With 50 Figures

Springer

Yasuo Ikeda, M.D., Dr. Med. Sci.
Professor, Department of Internal Medicine
Keio University School of Medicine
35 Shinanomachi, Shinjuku-ku, Tokyo 160–8582, Japan

Jun-ichi Hata, M.D., Dr. Med. Sci.
Professor, Department of Pathology
Keio University School of Medicine
35 Shinanomachi, Shinjuku-ku, Tokyo 160–8582, Japan

Shigeo Koyasu, Ph.D.
Professor, Department of Microbiology
Keio University School of Medicine
35 Shinanomachi, Shinjuku-ku, Tokyo 160–8582, Japan

Yutaka Kawakami, M.D., Dr. Med. Sci.
Professor, Institute for Advanced Medical Research
Keio University School of Medicine
35 Shinanomachi, Shinjuku-ku, Tokyo 160–8582, Japan

Yutaka Hattori, M.D., Dr. Med. Sci.
Department of Internal Medicine
Keio University School of Medicine
35 Shinanomachi, Shinjuku-ku, Tokyo 160–8582, Japan

ISBN-13: 978-4-431-68508-1   e-ISBN-13: 978-4-431-68506-7
DOI: 10.1007/978-4-431-68506-7

Library of Congress Cataloging-in-Publication Data

Cell therapy/Y. Ikeda ... [et al.] (eds.).
    p. ; cm.—(Keio University symposia for life science and medicine ; 5)
    Includes bibliographical references and indexes.
    ISBN-13: 978-4-431-68508-1
    1. Cellular therapy—Congresses. I. Ikeda, Yasuo, 1944– II. Keio International
Symposium for Life Sciences and Medicine (5th : 1999 : Keio University) III. Series.
    [DNLM: 1. Hematopoietic Stem Cell Transplantation—Congresses. 2. Hematopoietic
Stem Cells—physiology—Congresses. 3. Immunotherapy, Adoptive—Congresses. 4.
Neoplasms—therapy—Congresses. WH 380 C3925 2000]
    RM287 .C365 2000
    615.5—dc21
                                                                  00-025437
Printed on acid-free paper

SPIN: 10691861

Foreword

This volume contains the proceedings of the fifth symposium of the Keio University International Symposia for Life Sciences and Medicine under the sponsorship of the Keio University Medical Science Fund. As stated in the address by the Vice President of Keio University at the opening of the symposium, the fund was established by the generous donation of Dr. Mitsunada Sakaguchi. The Keio University International Symposia for Life Sciences and Medicine constitute one of the core activities of the fund. The objective is to contribute to the international community by developing human resources, promoting scientific knowledge, and encouraging mutual exchange. Every year, the Committee of the International Symposia for Life Sciences and Medicine selects the most interesting topics for the symposium from applications received in response to a call for papers to the Keio medical community. The publication of these proceedings is intended to publicize and distribute information arising from the lively discussions of the most exciting and current issues during the symposium. We are grateful to Dr. Mitsunada Sakaguchi, who made the symposium possible, the members of the program committee, and the office staff whose support guaranteed the success of the symposium. Finally, we thank Springer-Verlag, Tokyo, for their assistance in publishing this work.

Akimichi Kaneko, M.D., Ph.D.
Chairman
Committee of the International Symposia
for Life Sciences and Medicine
Keio University Medical Science Fund

Preface

This volume is the proceedings of the Keio University International Symposium for Life Sciences and Medicine on Cell Therapy, which was held at the Mita Campus of Keio University, January 21–23, 1999.

The symposium was intended to promote and contribute to the advancement of a new field of medicine by bringing together leading researchers and clinicians and providing an environment where active exchanges of important information and topics of research and clinical practice could take place.

Cell therapy is a new concept for the treatment of cancer, hematological malignancies, and autoimmune disorders, a concept that has progressed rapidly in recent years. The potential of this field to create advanced medicinal techniques capable of conquering refractory diseases is attracting attention not only among medical professionals but also among the general public. Recent advances in molecular biology and cell engineering technologies make it possible to apply the fruits of basic science to the treatment of disease. An example of the success of translational research is sure to appear soon in the field of cell therapy. Globalization of treatment methods is another important issue. In this area, the standardization of manipulated cells should be established in the near future.

The symposium was divided into three sessions. The first day ended with papers on immunotherapy for malignant diseases and immunization with tumor antigens. On the second day, the main focus was on hematopoietic stem cells. Various aspects of stem cells were discussed, including stem cell biology, regulatory issues, gene transfer to hematopoietic cells, and new strategies for hematopoietic stem cell transplantation. On the third day, papers on international collaboration in hematopoietic stem cell transplantation were presented, followed by a panel discussion.

We sincerely hope that this volume will enhance overall knowledge of cell therapy by serving as a reference and a source of inspiration. We wish to thank all participants for their contributions. We are particularly grateful to the administrative members of the Keio University Medical Science Fund, Hiroshi Ohin, Junko Shimane, and Sachiko Monden, for their excellent and dedicated help. Finally, we express our gratitude to the staff of Springer-Verlag, Tokyo, for their support.

Yasuo Ikeda
Jun-ichi Hata
Shigeo Koyasu
Yutaka Kawakami
Yutaka Hattori
EDITORS

Contents

Part 1 Immunotherapy for Malignant Diseases

Part 2 Hematopoietic Stem Cells Biology and Clinical Application

Part 3 International Collaboration in Hematopoietic Stem Cell Transplantation

Part 3 International Collaboration in Hematopoietic Stem Cell Transplantation

List of Contributors

Opening Remarks*

PROFESSOR RYUJI KOMATSU
VICE PRESIDENT, KEIO UNIVERSITY
MEMBER OF THE BOARD OF DIRECTORS
KEIO UNIVERSITY MEDICAL SCIENCE FUND

Distinguished Guests, Ladies and Gentlemen, and Dear Friends:

On behalf of Keio University, I wish to welcome all of you to this year's Keio University International Symposium for Life Sciences and Medicine.

It impresses me greatly that so many scholars and researchers have come not only from all over Japan but also from many other countries. We are particularly grateful to the guest speakers who kindly accepted our invitation despite their busy schedules. We feel reassured of your support in our endeavor to make further progress in life sciences and medicine, which is the objective of this symposium.

The first of the Keio University International Symposia for Life Sciences and Medicine was held in 1996 focusing on "Oxygen Homeostasis and Its Dynamics," and it is with great pleasure that I am here today to open the fifth of the series, the main topic of which is "Cell Therapy."

The 20th Century has been considered the era of economic development, and indeed, the capitalist economy has seen remarkable progress. Concurrently, many sciences have made equally impressive progress. In particular, the advances achieved in medicine have been outstanding. Such new fields as molecular biology have been developed, and researchers in cancer and leukemia, which are included in the topics of this year's symposium, have also taken big steps forward. This breathtaking progress has brought about, in significant measures, happiness, freedom, welfare, security, and, above all, increased hope for humankind.

Having said that, there still remain a number of unsolved obstacles in medicine and life sciences. Cures for cancer and leukemia are yet to be found and

---

* This opening address was given by Professor Ryuji Komatsu, Vice President of Keio University, at the opening session of the 5th Keio University International Symposium for Life Sciences and Medicine: Cell Therapy on the afternoon of Thursday, January 21, 1999, in the conference hall on the 2nd floor of the New North Building of the Mita Campus, Keio University

will rely greatly on future research and work. This means that higher and higher expectations are being placed on progress in this field in the 21st Century. I am convinced that all of you who are present here today are the very persons to respond to these expectations, and I am confident that this year's symposium will play a vital role in opening the door to further progress in the next century.

I now would like to talk briefly about Keio University. Keio is known as the first institution of higher education in Japan to teach economics. But also in the area of medicine, our university and its Mita Campus have been actively involved in promoting the modernization of Japan in various ways.

Yukichi Fukuzawa, who founded Keio University, is well known for having devoted his efforts to the advancement of medicine from the initial period of the Meiji Era. Although he himself was not a medical doctor, the most influential teacher he had was a medical doctor by the name of Koan Ogata. For that reason, Fukuzawa took a profound interest in medicine and, in later years, he himself came to give great support to promoting progress in medicine.

At the beginning of the Meiji Era, now about 125 years ago, Fukuzawa established a small medical school and a clinic here in Mita. His support for the research activities of Dr. Shibasaburo Kitasato is a well-known story. In this respect, it is only natural from the historical point of view that such a significant international symposium as this one should be held at Keio University.

Now, let me stress that this symposium would not be complete without mentioning the Keio University Medical Science Fund and the Sakaguchi Fund. In the fall of 1994, Dr. Mitsunada Sakaguchi, a member of the Class of 1940 of Keio Medical School, generously donated a sum of 5 billion yen to Keio University. The purpose of his donation was to encourage and facilitate original and high-level research activities in the field of life sciences and medicine, and thus to help advance the well-being of humankind. This symposium is organized as one of the many significant activities of the Fund. Dr. Sakaguchi is here today. Dr. Sakaguchi, I would like to take this opportunity to express again our most sincere appreciation for your immeasurable contribution. On this occasion I also wish to thank Dr. Yasuo Ikeda, whose endeavors and efficient organization as the leader of the whole project made this symposium possible.

Before I close, I would like to mention that being the oldest university in Japan, Keio owns two facilities on our hill at Mita which are important cultural assets of Japan. These are facilities that were involved, to a large extent, in the promotion of Japanese academia including medicine. I hope very much that you can find time to visit these sites while you are here.

Finally, let me conclude by extending my best wishes for the success of this symposium and, even more importantly, for the best of your health and further success in your meaningful work.

Thank you.

Part 1
Immunotherapy for Malignant Diseases

Breaking Immunological Tolerance to Tumor Cells as a Novel Immunotherapy for Cancer

Shimon Sakaguchi[1,2], Takeshi Takahashi[1,2], and Jun Shimizu[1]

Summary. The mechanisms that are responsible for maintaining immuno-logical tolerance to self-constituents may also act to impede immunity against autologous tumor cells. This review shows that one aspect of immunological self-tolerance in the periphery is maintained by T-cell-mediated control of self-reactive T cells, and that the same mechanism hampers the effective generation of tumor immunity as well. Furthermore, abrogation of this T-cell-mediated immunoregulation on tumor effector cells can indeed evoke potent immunity to autologous tumor cells in normal animals. These findings may help to devise a novel immunotherapy for cancer in humans.

Key words. Immunological self-tolerance, Autoimmunity, Tumor immunity, Regulatory T cells, CD25

Introduction

The mechanisms that maintain immunological tolerance to self-constituents may also act to impede immunity against autologous tumor cells. Indeed, many tumor antigens recognized by autologous cytotoxic T lymphocytes are antigenically normal self-constituents [1]. Furthermore, immunotherapy of cancer by vaccination with tumor antigens or infusion of ex vivo propagated cytotoxic lymphocytes often leads to the appearance of autoimmunity to normal tissues because of antigenic cross-reactions [2,3]. These findings suggest that manipulations of the immune system to break immunological

---

[1] Department of Immunopathology, Tokyo Metropolitan Institute of Gerontology, 35-2 Sakaecho, Itabashi-ku, Tokyo 173-0015, Japan
[2] Department of Experimental Pathology, Institute for Frontier Medical Sciences, Kyoto University, 53 Kawaharacho, Shogoin, Sakyo-ku, Kyoto 606-8397, Japan

self-tolerance might be able to evoke effective immune responses to autologous tumor cells as well.

In this chapter, we demonstrate that elimination of CD25 (interleukin-2 receptor α-chain) expressing T cells, which constitute 5%–10% of peripheral CD4+ T cells in normal mice and humans [4–9], not only breaks self-tolerance and causes autoimmune disease but also generates potent tumor-specific immune responses to syngeneic tumors and eradicates them. This novel way of evoking tumor immunity would contribute to devising effective immunotherapy for cancer in humans.

T-Cell-Mediated Maintenance of Immunological Self-Tolerance

In humans and animals, self-reactive T cells, especially CD4+ self-reactive T cells, play critical roles in the development of various autoimmune diseases by conducting cell-mediated immune responses against self-antigens or helping B cells to form autoantibody [10]. Clonal deletion of self-reactive T cells in the thymus and clonal anergy in the periphery are currently supposed to be the main mechanisms of controlling such self-reactive CD4+ T cells [11]. T-cell-mediated control of self-reactive T cells may also contribute to the maintenance of peripheral self-tolerance [12–14].

Although it is controversial which of these mechanisms is directly responsible for preventing autoimmune disease (in other words, whether autoimmune disease should develop when any of these mechanisms becomes aberrant), there are accumulating demonstrations that various autoimmune diseases can be produced in normal rodents by simply removing from the immune system a particular CD4+ T-cell subpopulation defined by expression levels of various cell-surface molecules [7–9,15–19] (e.g., CD5^{high}, CD45RB^{low}, or CD25+ CD4+ T cells) (Fig. 1). For example, when splenic cells suspensions prepared from normal BALB/c mice were depleted of CD25-expressing T cells and then transferred to T-cell-deficient BALB/c athymic nude mice, the recipients spontaneously developed various organ-specific autoimmune diseases (e.g., thyroiditis, gastritis, insulin-dependent diabetes mellitus, adrenalitis, sialoadenitis, and oophoritis) accompanying respective autoantibodies (e.g., autoantibodies specific for thyroglobulins, gastritc parietal cells, Langerhans islet cells, adrenocortical cells, the ductal cells of the salivary glands, or oocytes), and cotransfer of CD25+ CD4+ T cells with CD25– cells could prevent the autoimmune development [7,8] (Table 1).

The elimination of CD25+ CD4+ T cells also enhances immune responses to nonself antigens such as allogeneic tissue transplants [7]. Because the majority of CD25+ 4+ T cells are also CD5^{high} and CD45RB^{low}, it is likely that

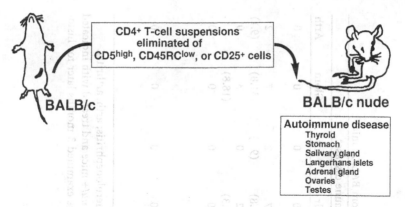

FIG. 1. Induction of autoimmune diseases in T-cell-deficient mice (or rats) by transferring a CD4+ T-cell subpopulation defined by the expression levels of various cell-surface molecules. The recipient animals spontaneously, without deliberate immunization with self-antigens, develop various autoimmune diseases immunopathologically similar to human counterparts

the immunoregulatory activity of CD5^{high} CD4+ cells or CD45RB^{low} CD4+ cells can be attributed to the CD25+ cells included in these populations (Fig. 2) [7]. Indeed, when the CD45RB^{low} CD4+ population in normal naive mice is dissected into CD25+ or CD25– cells, the CD25+ cells show potent suppressive activity on the proliferation of other CD4+ or CD8+ T cells in vitro, whereas the CD25– cells do not [20].

The CD25+ 4+ T cells in normal naive mice are unique in that they are non-proliferative (i.e., anergic) to in vitro antigenic stimulation, and that, upon stimulation, they exert potent suppression on the activation and proliferation of other T cells in an antigen-nonspecific manner; i.e., they suppress the proliferation of not only T cells with the same antigen specificity but also those specific for irrelevant antigens presented by the same antigen-presenting cells (APCs) [20]. The suppression is not mediated by far-reaching or long-lasting humoral factors including immunoregulatory cytokines such as IL-4, IL-10, or TGF-β, but dependent on cell-to-cell interactions on APCs [20,21]. These findings collectively indicate that one aspect of peripheral self-tolerance is maintained by CD25+ 4+ T-cell-mediated immunoregulation, and that abnormality of this mechanism may directly lead to the development of auto-immunity (Fig. 3).

Assuming that the CD25+ T-cell-mediated suppression is not antigen specific, a critical question is how it can control autoimmunity without impeding immune responses to nonself antigens. Given the fact that T cells expressing high-affinity T-cell receptors (TCRs) for intrathymic self-antigens or peripheral organ-specific antigens circulating to the thymus can be clonally

TABLE 1. Induction of autoimmune disease in BALB/c nude mice by inoculating T-cell subpopulations from BALB/c mice[a]

Group	Inoculated cells	Total number of mice	Number of mice with autoimmune disease[a]							
			Gas	Oop	Thyr	Sial	Adr	Ins	Glom	Arth
A	Whole (5×10⁷)	18	0	0	0	0	0	0	0	0
B	CD25− (5×10⁷)	22	22 (100)	22 (100)	16 (72.7)	10 (45.1)	7 (31.8)	2 (9.1)	7 (31.8)	2 (9.1)
C	CD4+ CD25− (5×10⁷)	16	14 (87.5)	13 (81.3)	7 (43.8)	5 (31.3)	2 (12.3)	0	3 (18.8)	0
D	CD8+ CD25− (5×10⁷)	10	0	0	0	0	0	0	0	0
E	+ CD25+ (2×10⁶)	6	1 (16.7)	0	0	0	0	0	0	0

Gas, gastritis; oop, oophoritis; thyr, thyroiditis; sial, sialoadenitis; adr, adrenalitis; ins, insulitis; glom, glomerulonephritis; arth, arthritis. Spleen and lymph node cells of indicated numbers were prepared from 2- to 3-month-old female BALB/c nu/+ mice and treated with indicated antisera and C, and then i.v. transferred to 6-week-old female nu/nu mice. The recipient nu/nu mice were examined 3 months later for histological and serological development of autoimmune diseases.
[a]Number of mice with histologically evident autoimmune disease (percent incidence in parentheses).

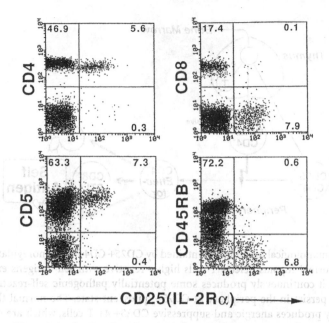

FIG. 2. Expression of the CD25 molecule on a subpopulation of peripheral CD4+ T cells in normal naive mice. Lymph node cells from 2-month-old normal BALB/c mice are stained with fluorescent isothiocyanate (FITC)-labeled anti-CD25 antibody (*abscissa*) and various phycoerythrin (PE)-labeled antibody (*ordinate*), and analyzed by cytofluorometry

deleted even at a much lower antigen concentration than that required for activation [22], it is likely that self-reactive T cells having escaped the thymic negative selection may bear lower-affinity TCRs for the relevant self-antigens, hence higher thresholds for activation, than T cells specific for nonself antigens. Thus, the CD25+ T-cell-mediated immunoregulation, which appears to upregulate the activation thresholds of dormant T cells in an antigen-nonspecific manner, may be sufficient for controlling these low-affinity self-reactive T cells; by contrast, high-affinity T cells specific for nonself antigens may easily overwhelm the regulation and become activated upon antigen presentation. The T-cell-mediated immunoregulation and the thymic clonal deletion may thus cooperatively contribute to sustaining stable self-tolerance while enabling effective immune responses to nonself antigens (Fig. 4).

Induction of Tumor Immunity by Manipulating a T-Cell Subpopulation

These findings on self-tolerance and autoimmunity suggest the possibility that elimination of CD25+ T cells may evoke potent tumor-specific immune responses to syngeneic tumors as quasi-autoimmune responses [23]

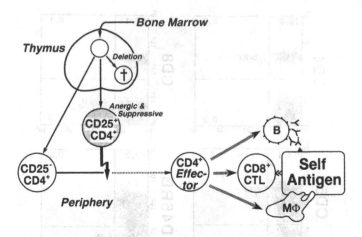

FIG. 3. Immmunological tolerance maintained by CD25+ CD4+ immunoregulatory T cells. While the normal thymus deletes T cells highly reactive with self-antigens expressed in the thymus, it continuously produces some potentially pathogenic self-reactive CD4+ T cells, which persist in the periphery at CD25– quiescent state. The normal thymus also continuously produces anergic and suppressive CD25+ 4+ T cells, which are suppressing the activation and expansion of CD4+ self-reactive T cells from the CD25– dormant state. When the immunoregulatory CD25+ 4+ T cells are eliminated or substantially reduced, or their regulatory function is impaired, CD25– self-reactive T cells become activated, expand, and differentiate to CD25+-activated autoimmune effector T cells (*dotted arrow*), which may help B cells to form autoantibodies (Th2 response), conduct cell-mediated immune responses by recruiting inflammatory cells, including activated macrophages (MΦ) (Th1 response), or help activation and expansion of CD8+ cytotoxic lymphocytes (CTLs)

(Fig. 4). To examine this possibility, we transferred to BALB/c athymic nude mice BALB/c splenic cell suspensions depleted of CD25+ cells, and then subcutaneously transplanted BALB/c-derived RL♂1 leukemia cells [24,25] (Fig. 5). In the majority of mice, the tumors first grew and then regressed within a month, allowing the hosts to survive a long term (>80 days), whereas all the nude mice transferred with nondepleted spleen cells or the mixture of an equal number of CD25– cells and CD4+ T cells died of tumor progression within 40 days. Upon rechallenge with larger doses of RL♂1, the CD25– cell-transferred nude mice rejected the tumor cells more rapidly and vigorously than the primary rejection, indicating that they had become immune to the tumor. These mice indeed developed tumor-specific cytotoxic T lymphocytes (CTLs).

The CD25– cell-transferred nude mice that had rejected tumor transplants (as shown in Fig. 5) also developed histologically evident autoimmune diseases including gastritis (100% incidence; $n = 14$), thyroiditis (42.9%), and oophoritis (64.3%), accompanying respective autoantibodies, as did the nude

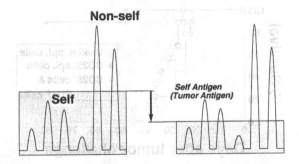

FIG. 4. Immune responsiveness to self and nonself depicted as a continuum. The *horizontal line* indicates the level of immunoregulation by CD25+ T cells. The *peaks* above the line represent overt immune responses to self- or nonself antigens. When the level goes down (i.e., the immunoregulation becomes weaker), immune responses to certain self-antigens become apparent while responses to nonself antigens become enhanced, presumably because the activation thresholds of T cells become lowered. The peaks representing immune responsiveness to self-antigens are depicted as being lower than those to nonself antigens, because the thymic clonal deletion mechanism supposedly deletes T cells bearing high-avidity T-cell receptors (TCRs) for the self-antigens expressed in the thymus or expressed in the periphery but circulating to the thymus, even in a minute quantity

mice transferred with CD25– cells (but not tumor-transplanted), as shown in Table 1. Thus, elimination of CD25+ CD4+ immunoregulatory T cells not only elicits autoimmunity but also generates potent tumor immunity against syngeneic tumor cells.

CD25+ CD4+ Immunoregulatory T Cells in Autoimmunity and Tumor Immunity

It has been postulated that one of the elements impeding effective tumor immunity in tumor-bearing hosts may be concomitant development of a T-cell population suppressing the generation or action of tumor-killing effector cells [26–28]. Although some of such suppressor T cells were shown to be CD4+, they have eluded further characterization and manipulation because of the lack of reliable markers specific for them [29]. Our results indicate that these suppressive T cells, at least in part, can be CD25+ 4+ T cells. The CD25+ 4+ immunoregulatory T cells, however, bear several characteristics distinct from the suppressor T cells concomitantly induced by sensitization to tumor antigens [27–29].

First, removal of CD25+ 4+ T cells before tumor implantation was effective in evoking specific tumor immunity. Second, their removal elicited not only tumor immunity but also various autoimmunities (Table 1; Fig. 5). Third, such

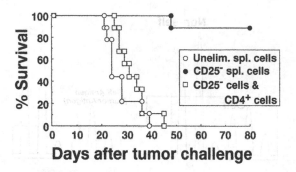

FIG. 5. Eradication of tumor transplants in nude mice by transferring CD25+ cell-depleted splenic cell (CD25– cell) suspensions: nude mice transferred with CD25– spleen cells (*closed circles*), untreated spleen cells (*open circles*), or the mixtures of CD25– cells and CD4+ cells (*open squares*). Individual BALB/c nude mice were subcutaneously transplanted with 1.5×10^5 RL♂1 cells immediately after intravenous transfer of 3×10^7 untreated spleen cells, 3×10^7 CD25– spleen cells, or mixtures of the same number (3×10^7) of CD25– spleen cells and CD4+ spleen cells. Percentage of mice surviving in each group ($n = 9$) on various days after tumor transplantation (total of three independent experiments) is shown

immunoregulatory CD25+ 4+ T cells are continuously produced by the normal thymus and already functional in the thymus [21]. These findings when taken together indicate that the CD25+ 4+ immunoregulatory T cells present in normal naive mice may form a functionally and developmentally distinct T-cell subpopulation, which may be continuously exerting suppression by raising the activation thresholds of other T cells in an antigen-nonspecific manner, thereby impeding effective generation of tumor immunity and inhibiting autoimmune responses. To further elucidate how the CD25+ 4+ immunoregulatory T cells suppress tumor immunity (and autoimmunity), the natural ligands that they recognize must be determined, as they require TCR stimulation to exert the suppressive activity [20,21]. Their "activated" phenotype (e.g., being CD25+, CD45RB^{low}, CD44^{high}, and CD5^{high} [7,8,20,21]) in normal naive mice suggests that the CD25+ 4+ immunoregulatory T cells might be inherently reactive with self-antigens and continuously activated by them in the normal internal environment, although they themselves may be harmless because of their anergic property [20,21].

Removal of CD25+ 4+ T cells elicited autoimmunity in addition to tumor rejection. This result raises the issue how tumor immunity can be evoked without autoimmunity upon elimination of CD25+ 4+ T cells. We have previously shown that the intensity and the range of autoimmune responses (i.e., the severity, the incidence, and the spectrum of autoimmune diseases) elicited by removing the immunoregulatory T cells depend on the T-cell subpopulations involved, the degree and duration of depleting CD25+ T cells, and the genetic background of the hosts [7,23]. For example, in genetically autoimmune-prone BALB/c mice, generation of effective tumor immunity can be achieved without deleterious autoimmunity by limiting the period of depleting CD25+ T cells by in vivo administration of anti-CD25 monoclonal antibody for a limited period (S. Yamazaki, J. Shimizu, and S. Sakaguchi, manuscript in preparation). On the other hand, in genetically autoimmune-resistant C57BL/6 mice, complete depletion of CD25+ 4+ T cells from cell inocula (see Fig. 5) leads to tumor rejection but fails to cause autoimmune disease in syngeneic nude mice (S. Yamazaki, J. Shimizu, and S. Sakaguchi, unpublished results). Thus, autoimmunity and tumor immunity elicited by abrogation of the CD25+ 4+ T-cell-mediated immunoregulation can be somehow differentiated by the duration or degree of CD25+ T-cell depletion required for induction of autoimmunity or tumor immunity, or by genetic makeup of the hosts (i.e., genetically determined susceptibility or resistance to autoimmune disease by the hosts).

Conclusions and Perspectives

The present finding that elimination of CD25+ CD4+ T cells can evoke effective tumor immunity can be instrumental in devising a novel cancer immunotherapy, with a caveat of possible autoimmunity. For example, administration of anti-CD25 antibody for a limited period to eliminate CD25+ 4+ immunoregulatory T cells may be effective in humans as well. Furthermore, in vitro culture of CD25− spleen cells from tumor-unsensitized mice can generate tumor effector cells including CTLs and NK cells (J. Shimizu and S. Sakaguchi, manuscript submitted). Transfusion of such tumor-killing effector lymphocytes induced in vitro by culturing peripheral blood CD25− lymphocytes or CD25− tumor-infiltrating lymphocytes can be effective in tumor rejection when combined with subsequent in vivo IL-2 administration [30].

Acknowledgment. This work was supported by grants-in-aid from the Ministry of Education, Science, Sports and Culture, the Ministry of Human Welfare, and the Organization for Pharmaceutical Safety and Research (OPSR) of Japan.

References

1. Boon T, Cerrottini JC, Van den Eynde B, van der Bruggen P, Van Pel A (1994) Tumor antigens recognized by T lymphocytes. Annu Rev Immunol 12:337–365
2. Rosenberg SA, White DE (1996) Vitiligo in patients with melanoma: normal tissue antigens can be targets for cancer immunotherapy. J Immunother 19:81–84
3. Naftzger C, Takechi Y, Kohda H, Hara I, Vijayasaradhi S, Houghton AN (1996) Immune

response to a differentiation antigen induced by altered antigen: a study of tumor rejection and autoimmunity. Proc Natl Acad Sci USA 93:14809–14814

4. Uchiyama T, Broder S, Waldmann TA (1981) A monoclonal antibody (anti-Tac) reactive with activated and functionally mature human T cells. I. Production of anti-Tac monoclonal antibody and distribution of Tac (+) cells. J Immunol 126:1393–1397

5. Jackson AL, et al (1990) Restricted expression of p55 interleukin 2 receptor (CD25) on human T cells. Clin Immunol Immunopathol 54:126–133

6. Kanegane H, et al (1991) A novel subpopulation of CD45RA+ CD4+ T cells expressing IL-2 receptor α-chain (CD25) and having a functionally transient nature into memory cells. Int Immunol 3:1349–1356

7. Sakaguchi S, et al (1995) Immunologic tolerance maintained by activated T cells expressing IL-2 receptor α-chains (CD25): breakdown of a single mechanism of self-tolerance causes various autoimmune diseases. J Immunol 155:1151–1164

8. Asano M, Toda M, Sakaguchi N, Sakaguchi S (1996) Autoimmune disease as a consequence of developmental abnormality of a T cell subpopulation. J Exp Med 184: 387–396

9. Suri-Payer E, Amar AZ, Thornton AM, Shevach EM (1998) CD4+ CD25+ T cells inhibit both the induction and effector function of autoreactive T cells and represent a unique lineage of immunoregulatory cells. J Immunol 160:1212–1218

10. Schwartz RS (1993) Autoimmunity and autoimmune diseases. In: Paul WE (ed) Fundamental immunology, 3rd edn. Raven Press, New York, pp 1033–1097

11. Sinha AA, Lopez MT, McDevitt HO (1990) Autoimmune diseases: the failure of self tolerance. Science 248:1380–1388

12. Modigliani Y, Bandeira A, Coutinho A (1996) A model of developmentally acquired thymus-dependent tolerance to central and peripheral antigens. Immunol Rev 149:155–175

13. Fowell D, McKnight AJ, Powrie F, Dyke R, Mason D (1991) Subsets of CD4+ T cells and their roles in the induction and prevention of autoimmunity. Immunol Rev 123:37–64

14. Sakaguchi S, Sakaguchi N (1994) Thymus, T cells and autoimmunity: various causes but a common mechanism of autoimmune disease. In: Coutinho A, Kazatchkine M (eds) Autoimmunity: physiology and disease. Wiley-Liss, New York, pp 203–227

15. Sakaguchi S, Fukuma K, Kuribayashi K, Masuda T (1985) Organ-specific autoimmune diseases induced in mice by elimination of T-cell subset. I. Evidence for the active participation of T cells in natural self-tolerance: deficit of a T-cell subset as a possible cause of autoimmune disease. J Exp Med 161:72–87

16. Sugihara S, et al (1988) Autoimmune thyroiditis induced in mice depleted of particular T-cell subset. I. Requirement of Lyt-1^{dull} L3T4^{bright} normal T cells for the induction of thyroiditis. J Immunol 141:105–113

17. Smith H, Lou YH, Lacy P, Tung KSK (1992) Tolerance mechanism in experimental ovarian and gastric autoimmune disease. J Immunol 149:2212–2218

18. Powrie F, Mason D (1990) OX-22high CD4+ T cells induce wasting disease with multiple organ pathology: prevention by OX-22low subset. J Exp Med 172:1701–1708

19. McKeever U, et al (1990) Adoptive transfer of autoimmune diabetes and thyroiditis to athymic rats. Proc Natl Acad Sci USA 87:7618–7622

20. Takahashi T, et al (1998) Immunologic self-tolerance maintained by CD25+ CD4+ naturally anergic and suppressive T cells; induction of autoimmune disease by breaking their anergic/suppressive state. Int Immunol 10:1169–1180

21. Itoh M, Takahashi T, Sakaguchi N, Kuniyasu Y, Shimizu J, Otsuka F, Sakaguchi S (1999) Thymus and autoimmunity: production of CD25+ CD4+ naturally anergic and sup-

pressive T cells as a key function of the thymus in maintaining immunologic self-tolerance. J Immunol 162:5317–5326

22. Pircher H, Rohrer UH, Moskophides D, Zinkernagel RM, Hengartner H (1991) Lower receptor avidity required for thymic clonal deletion than for effector T-cell function. Nature (Lond) 351:482–485

23. Sakaguchi S, et al (1996) T-cell mediated maintenance of self-tolerance and its breakdown as a possible cause of various autoimmune diseases. J Autoimmun 9:211–220

24. Nakayama E, et al (1979) Definition of a unique cell surface antigen of mouse leukemia RL♂1 by cell-mediated cytotoxicity. Proc Natl Acad Sci USA 76:3486–3490

25. Kuribayashi K, Keyaki A, Sakaguchi S, Masuda T (1985) Effector mechanisms of syngeneic anti-tumor responses in mice. Immunology 56:127–140

26. Fujimoto S, Greene MI, Sehon AH (1976) Regulation of the immune response to tumor antigen. I. Immunosuppressor cells in tumor-bearing hosts. J Immunol 116:791–799

27. Naor D (1979) Suppressor cells: permitters and promotors of malignancy. Adv Cancer Res 29:45–125

28. North RJ (1982) Cyclophosphamide-facilitated adoptive immunotherapy of an established tumor depends on elimination of tumor-induced suppressor T cells. J Exp Med 55:1063–1074

29. Awwad M, North RJ (1988) Immunologically mediated regression of a murine lymphoma after treatment with anti-L3T4 antibody: a consequence of removing L3T4+ suppressor T cells from a host generating predominantly Lyt-2+ T cell-mediated immunity. J Exp Med 168:2193–2206

30. Rosenberg SA, Lotze MT (1986) Cancer immunotherapy using interleukin-2 and interleukin-2-activated lymphocytes. Annu Rev Immunol 4:681–710

SART1 Gene Encoding Squamous Cell Carcinoma Antigen Recognized by Cytotoxic T Lymphocytes

Kyogo Itoh[1], Shigeki Shichijo[1], Yoshiko Inoue[2], Akihiro Hayashi[3], Uhi Toh[3], and Hideaki Yamana[3]

Summary. With the exception of melanomas, tumor antigens recognized by cytotoxic T lymphocytes (CTLs) are yet unidentified. We have identified the *SART1* gene encoding tumor epitopes of human squamous cell carcinomas (SCCs) recognized by HLA-A2601-restricted CTLs. This gene showed no similarity to known sequences, and encoded two proteins, 125 kDa of the $SART1_{800}$ nuclear protein and 43 kDa of the $SART1_{259}$ cytosol protein. The $SART1_{800}$ protein with leucine-zipper motif was expressed in the nucleus of the majority of proliferating cells tested including normal and malignant cells, whereas the $SART1_{259}$ protein was expressed in the cytosol of most SCCs from various organs and one-third to one-half of adenocarcinomas, but was not expressed in melanomas or leukemias or in a panel of normal tissues except for testis. The SART1 peptide at position 736–744 located in the region shared by the two proteins was recognized by the HLA-A26-restricted CTLs, and was able to induce in vitro the HLA-A26-restricted and $SART-1_{259}^+$ tumor-specific CTLs from peripheral blood mononuclear cells (PBMCs) of cancer patients. On the other hand, the SART1 peptide at position 690–698 was recognized by the HLA-A24-restricted CTLs, and also had the ability to induce the HLA-A24-restricted CTLs recognizing the $SART-1_{259}^+$ tumor cells in PBMCs of HLA-A24 cancer patients. Therefore, The $SART-1_{259}$ protein and its peptides may be useful for specific immunotherapy of epithelial cancer patients.

Key words. Tumor antigen, Cytotoxic T lymphocytes, Peptide antigen, Epithelial cancer, Cancer vaccine

[1]Departments of Immunology and [3]Surgery, Kurume University School of Medicine, [2]Cancer Vaccine Division of Kurume University Research Center for Innovative Cancer Therapy, 67 Asahi-machi, Kurume, Fukuoka 830-0011, Japan

Introduction

Many genes encoding tumor-rejection antigens recognized by cytotoxic T lymphocytes (CTLs) were identified from cDNA of melanomas [1–4]. A large number of nonapeptides recognized by HLA-class I-restricted CTLs cytotoxic to melanoma cells have been identified in the past 5 years [4–7]. Some of them are under clinical trials with major tumor regression in melanoma patients [8–10]. Therefore, these nonapeptides recognized by the CTLs could be a new tool for the specific immunotherapy of melanoma. However, no peptides are yet identified from human squamous cell carcinomas (SCCs) except for a mutated CASP-8 [11] and an ESO-1 that was identified by the serological approach [12]. This chapter reviews the recently identified gene (SART1) encoding the two novel proteins and the nonapeptides as the antigens recognized by the HLA-A26-restricted CTLs [13] and the HLA-A24-restricted CTLs [14].

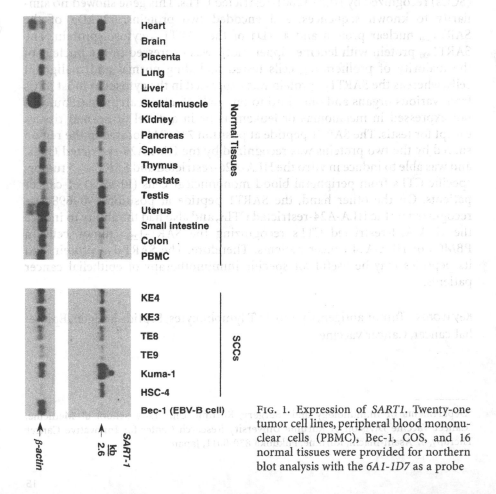

FIG. 1. Expression of SART1. Twenty-one tumor cell lines, peripheral blood mononuclear cells (PBMC), Bec-1, COS, and 16 normal tissues were provided for northern blot analysis with the 6A1-1D7 as a probe

Identification of the *SART1* Gene

Expression-gene cloning methods developed by Boon and his colleagues [4] were employed in this study to identify a gene coding tumor antigen recognized by the KE4 CTLs [14]. This method allows the identification of genes encoding tumor-rejection antigens [1–6]. After repeated experiments, one clone (*6A1-1D7*) was confirmed to encode a tumor antigen recognized by the KE4 CTLs. The sequence of this cDNA clone proved to be 990 bp long. However, by northern blot analysis, a band of about 2.6 kb was observed in all the normal tissues and tumor cell lines tested (Fig. 1). The relative level of mRNA expression was within the range of 2.3 ± 2.9 in all the samples except for testis (expression level, 7.5) and pancreas (17.4). These results suggest that this gene was ubiquitously expressed at the mRNA level with higher expression in testis and pancreas, and that the 990-bp-long cDNA was incomplete.

Subsequently, a 2506-bp-long gene was independently cloned from the cDNA libraries of KE4 tumor and PBMCs of healthy donors using the *6A1-1D7* as a probe. Both clones contained the *6A1-1D7* at positions 1517 to 2506. This 2506-bp-long gene showed no similarity to known sequences, and was tentatively termed the *SART1* (*Squamous cell carcinoma antigen recognized by T cells 1*) gene. The sequence data of the *SART1* are available from EMBL/GebBank/DDBJ under accession number AB006198. Although the *SART1* mRNA was ubiquitously expressed, the KE4 CTLs did not recognize nonmalignant cells including Bec-1 cells [15], possibly because of preferential expression of tumor antigens on the malignant cells by the mechanism of posttranscriptional regulation.

Expression of *SART1* Antigens

A polyclonal antibody to the $SART1_{800/GST}$ recognized an 125-kDa band of $SART1_{800}$ protein after cleavage of glutathione-s-transferase (GST) with thrombin (data not shown), and recognized an 125-kDa band in the nuclear fraction of peripheral blood mononuclear cells (PBMCs) activated with 10 μg/ml of phytohemagglutm (PHA) (PHA blasts), KE4 tumor, and Bec-1, but not unstimulated PBMCs (Fig. 2a). No protein in the cytosol was recognized by this Ab in any sample tested. The 125-kDa band was also expressed in the nucleus of the majority of tumor tissues, tumor cell lines, and normal cell lines tested, but was not expressed in normal tissues except for testis and fetal liver (Table 1). When the *SART1* of positions 29 to 2449 ($SART1_{800}$) was transfected to VA13, intensities of the 125-kDa band in both the nuclear and cytosol fractions increased (Fig. 2a). Furthermore, this Ab and anti-myc monoclonal Ab recognized a 132-kDa band of the COS cells transfected with the *SART1* of positions 29 to 2449 in conjunction with *pcDNA3.1/Myc-His* vector

18   K. Itoh et al.

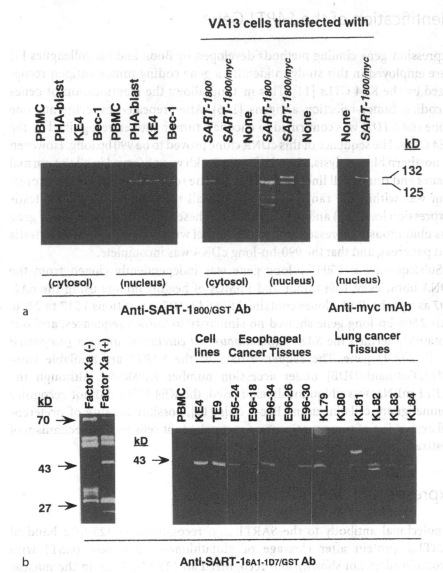

FIG. 2a,b. Expression of SART1$_{800}$ and SART1$_{259}$ proteins. Tumor cell lines used for western blot analysis were head and neck squamous cell carcinomas (SCCs), esophageal SCCs, lung adenocarcinomas, and lung SCC, leukemia cells, and melanomas. PBMCs, PHA blasts, fibroblast cells, and tumor tissues from various organs were also studied. a Expression of the SART1$_{800}$ protein was investigated by western blot analysis with anti-SART1$_{800/GST}$ Ab. Anti-myc monoclonal Ab (Invitrogen) was also used for analysis of VA13 transfected with the SART1$_{800/myc}$. b Expression of the SART1$_{259}$ was investigated with anti-SART1$_{6A1-1D7/GST}$ Ab. In the *left* gel, 70-, 43-, and 27-kDa bands correspond to the SART1$_{6A1-1D7/GST}$, SART1$_{6A1-1D7}$, and GST, respectively. Data for the cytosol fraction are shown

TABLE 1. Expression of the $SART1_{800}$ and $SART1_{259}$ proteins in normal and cancer cells and tissues

	$SART1_{800}$ (nucleus)[a]		$SART1_{259}$ (cytosol)[b]	
	Cell lines	Tissues	Cell lines	Tissues
Normal cells				
PBMC	0/5[c]		0/5	
PHA blast	2/2		0/2	
Fibroblast	2/2		0/2	
Fetal liver		1/1		1/1
Newborn liver		0/1		0/1
Liver		0/1		0/1
Testis		1/1		3/3
Placenta		0/1		0/2
Esophagus		0/2		0/4
Pancreas		0/1		0/1
Cancer cells				
Head and neck SCC	2/2	2/2	3/5	7/7 (100%)
Esophageal SCC	5/5	3/5	4/6	18/30 (60%)
Lung cancer				
Adenocarcinoma	3/3	7/7	3/6	16/35 (47%)
SCC	2/2	3/4	3/3	8/17 (47%)
Leukemia	4/4	4/4	0/16	0/10 (0%)
Melanoma	1/1		0/2	0/10 (0%)

PBMC, peripheral blood mononuclear cells; SCC, squamous cell carcinoma.
[a] Expression of the $SART1_{800}$ protein in the nucleus of various normal and cancer cells and tissues was investigated by western blot analysis with anti-$SART1_{800/GST}$ Ab.
[b] Expression of the $SART1_{259}$ protein in the cytosol of various normal and cancer cells and tissues was investigated by western blot analysis with anti-$SART1_{6A1-1D7/GST}$ Ab.
[c] Number of positive per total samples tested.

($SART1_{800/myc}$). The different migration of these bands (125 and 132-kDa) is caused by a tag peptide (theoretically about 5 kDa). These results suggest that the 125 kDa of the $SART1_{800}$ protein was expressed in the nucleus of proliferating cells, including normal and malignant cells, but not in nonproliferating cells or any normal tissues except for testis and fetal liver.

An Ab to the $SART1_{6A1-1D7/GST}$ recognized a 43-kDa band of the recombinant $SART1_{6A1-1D7}$ protein after cleavage of GST with Factor Xa (Fig. 2b). Therefore, the $SART1_{259}$ could be translated by the mechanism of −1 frameshifting in the prokaryotic mRNA and recognized by anti-$SART1_{6A1-1D7/GST}$ Ab. This Ab also recognized a 43-kDa protein in the cytosol of KE4 and TE9 esophageal SCC cell lines, fresh esophageal SCCs, and lung SCCs and adenocarcinomas, but not PBMCs (Fig. 2b). No protein in the nucleus was detected by this Ab in any sample tested (data not shown). The 43-kDa protein was expressed in the cytosol of all the head and neck SCC tissues tested, 60% of esophageal SCCs, and half of the lung SCCs and lung adenocarcinomas, but was not observed in leukemia, melanomas, or any normal tissues, normal cell lines, or normal cells except for fetal liver and testis (Table 1).

All the results suggest that the *SART1* gene is a bicistronic gene encoding two proteins, the 125kDa of $SART1_{800}$ in the nucleus and the 43kDa of $SART1_{259}$ in the cytosol. Most eukaryotic mRNAs have a single ORF and a single functional initiation site, which is usually the AUG codon that lies closest to the 5'-end [16,17]. However, there are some viral mRNAs that break this rule in which two proteins are translated from either the same or different ORF [18]. Several human genes are also suggested to be bicistronic [19]. *LAP* mRNA was found to be translated into two proteins, LAP and the liver-enriched transcriptional-inhibitory protein (LIP) [20]. LIP contains the DNA-binding and dimerization domains but is devoid of the transcription-activation domain; LAP and LIP seem to exhibit antagonistic activities. Another example is the *gp75* encoding two different polypeptides, gp75 recognized by sera from cancer patients and a peptide with 24 amino acids recognized by CTLs [19]. However, the mechanism of posttranscriptional regulation in human mRNAs is little understood at the present time. Therefore, *SART1* should be a novel tool to explore the mechanism.

Antigenic Epitopes Recognized by the HLA-26-Restricted CTLs

Data from the experiments of deletion mutants of both the *6A1-1D7* and *SART1* genes indicated that antigenic peptide(s) resided within the 186-bp region of the 3'-end of both the *6A1-1D7* and *SART1* [13]. This region encodes 62 deduced amino acids of the SART1 protein at positions of 737 to 800 in the third frame, which was shared by both the *6A1-1D7*-derived $SART1_{259}$ protein and *SART1*-derived $SART1_{800}$ protein. Subsequently, a series of 22 SART1 oligopeptides (10-mers) corresponding to the region shown were loaded to the VA13 cells that had been transfected with *HLA-A2601* or *-A0201*, and tested for their ability both to stimulate IFN-γ production and to be recognized by the KE4 CTLs in a ^{51}Cr-release assay (Fig. 3). The three 10-mers ($SART1_{736-745}$ <KGSGKMKTER>, $SART1_{748-757}$ <KKLDEEALLK>, $SART1_{784-793}$<IVLSGSGKSM>) possessed the activity to stimulate significant levels of IFN-γ production (>50pg/ml), whereas none of the other 10-mers did so. The $SART1_{736-745}$ or $SART1_{748-757}$ peptide had high (45% lysis at an E:T ratio of 5:1) or low (10% lysis) activity recognized when loaded on VA13 cells transfected with *HLA-A2601*, respectively. None of the other 10-mers, including the $SART1_{784-793}$, had a significant level (>10% lysis) of activity in a ^{51}Cr-release assay. Six different nonapeptides from these three 10-mers with deletion of one amino acids at position 1 or 10 were tested

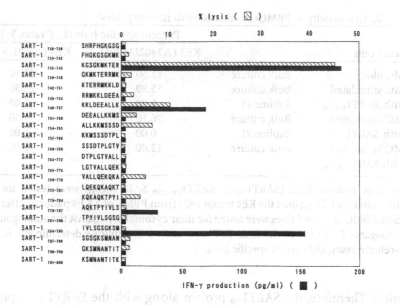

FIG. 3. Determination of peptide antigens. A series of 22 SART1 oligopeptides (10-mers) (10 nM) were loaded for 2 h to the VA13 cells (2×10^4) transfected with *HLA-A2601* or -*A0201*. For IFN-γ production, the KE4 CTLs (1×10^4) were added and incubated for 18 h; the culture supernatant was collected for measurement of IFN-γ by ELISA in duplicate assays. In a 6-h ^{51}Cr-release assay, these VA13 cells were labeled with $Na_2{}^{51}CrO_4$ for 1 h followed by addition of the KE4 CTLs (5×10^4). *Striped bars* indicate % lysis

for their ability to stimulate IFN-γ production by the parental KE4 CTL. Each nonapeptide (SART1$_{736-744}$ <KGSGKMKTE>, SART1$_{749-757}$ <KLDEE-ALLK>, SART1$_{785-793}$<VLSGSGKSM>) had higher activity to stimulate IFN-γ production than had the parental 10-mer (data not shown). In contrast, all the remaining nonapeptides failed to stimulate IFN-γ production.

The three nonapeptides (SART1$_{736-744}$, SART1$_{749-757}$, SART1$_{785-793}$) were tested for their ability to induce CTLs against the autologous tumor cells from PBMCs of a KE4 patient. PBMCs stimulated with the SART1$_{736-744}$ and their subline #1 showed higher levels of the KE4 autologous tumor cell lysis as compared to those of the KE3 allogenic tumor cell lysis (Table 2). In contrast, PBMCs cultured with IL-2 alone or stimulated with SART1$_{749-757}$ or SART1$_{785-793}$ equally lysed both tumor cells. The SART1$_{736-744}$ peptide, but not the others, induced from a patient's PBMCs the CTLs restricted to the autologous tumor cells. The HLA-A26 allele is found in approximately 22% of Japanese, 17% of Caucasians, and 14% of Africans [21,22]. The A2601 subtype is found most frequently among the A26 subtypes [22,23]. Further, SART1$_{736-744}$ peptide has the ability to induce HLA-A26-restricted CTLs in HLA-A2601, -2602, and -2603 cancer patients (Inoue et al., unpublished

TABLE 2. Cytotoxicity in PBMCs stimulated with nonapeptides[a]

Effector cells		Percent specific lysis (E:T ratio, 5:1)	
		KE3 (A2402/A0201)	KE4 (A2601/A2402)
PBMC alone	Bulk culture	11.50	12.10
PBMC stimulated	Bulk culture	15.80	28.50
with SART1$_{736-744}$	Subline #1	0.00	12.00
PBMC stimulated	Bulk culture	26.10	27.50
with SART1$_{749-757}$	Subline #1	0.00	0.00
PBMC stimulated	Bulk culture	12.00	13.00
with SART1$_{785-793}$			

[a] The three nonapeptides (SART1$_{736-744}$, SART1$_{749-757}$, SART1$_{785-793}$) were tested for their ability to induce CTL against the KE4 tumor cells from PBMC of a KE4 patients. After stimulation, PBMC or the sublines were tested for their cytotoxicity against the autologous KE4 and allogenic KE3 tumor cells at an E:T ratio of 5:1 in triplicate determinants in a 6-h ^{51}Cr-release assay. Data are % specific lysis.

results). Therefore, the SART1$_{259}$ protein along with the SART1$_{736-744}$ peptide could be useful for specific immunotherapy of relatively large numbers of HLA-A26 patients with SCCs or adenocarcinomas as cancer vaccine, and also as an antigen in vitro to induce CTLs for adoptive cellular therapy.

Antigenic Epitopes Recognized by the HLA-24-Restricted CTLs

An HLA-A24 allele is found in 60% of Japanese [22]. The SART1 antigen possesses many peptide sequences with an HLA-A24-binding motif. Therefore, we have investigated if these HLA-A24-binding peptides can be available for specific immunotherapy of HLA-A24$^+$ SCC and adenocarcinoma patients. For these experiments, the HLA-A24-restricted and SART1$_{259}$-tumor-specific CTL line [14] was used. We initially tested the reactivity of this CTL line to the 18 different peptides with HLA-A24-binding motifs that were originated from SART1$_{800}$ and SART1$_{259}$ antigen (Fig. 4). The CTL line produced significant levels (>400pg/ml) of IFN-γ by recognition of HLA-A24-transfected VA13 cells that had been pulsed with the SART1$_{690-698}$, SART1$_{746-755}$, or SART1$_{758-766}$. These three peptides are expected to be shared by both SART1$_{800}$ and SART1$_{259}$. The highest level of IFN-γ production was consistently observed in the CTLs by recognition of the SART1$_{690-698}$ (EYRGFTQDF) in the different experiments. The results suggest this peptide (EYRGFTQDF) is mainly recognized by the HLA-A24-restricted and SART1-specific CTLs.

These three peptides recognized by the CTL line were further tested for the ability to induce CTLs recognizing HLA-A24$^+$ tumor cells from PBMCs of HLA-A24 cancer patients and healthy donors. PBMCs that had been stimulated in vitro three times with the SART1$_{690-698}$ produced higher levels of

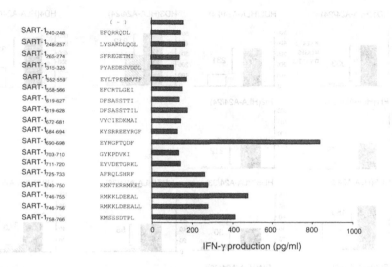

FIG. 4. Recognition of peptides. The HLA-A24-restricted CTLs were tested for their activity to produce IFN-γ by recognition of the *HLA-A2402*-transfected VA13 cells that had been pulsed with one of the 18 different peptides, or the *HLA-A2601*-transfected VA13 cells with some of these peptides as a negative control. The peptide (10 μM) was loaded for 2 h to VA13 cells (1 × 10^4) transfected with *HLA-A2402* or *-A2601* followed by testing their ability to stimulate IFN-γ production by the CTLs. For IFN-γ production, the CTLs (2 × 10^4) were added, incubated for 18 h in triplicate assays, and the culture supernatant collected for measurement of IFN-γ by ELISA

IFN-γ by recognition of the KE4 tumor (HLA-A24$^+$, SART1$_{259}$$^+$, and SART1$_{800}$$^+$), but not QG56 (HLA-A24$^-$, SART1$_{259}$$^+$, and SART1$_{800}$$^+$) tumor cells or VA13 (HLA-A24$^-$, SART1$_{259}$$^-$, and SART1$_{800}$$^+$) fibroblast cells in all the six HLA-A24 homozygotes ($n = 4$ in healthy donors [HDs] and $n = 2$ in cancer patients [Pts]) tested (Fig. 5). Levels of IFN-γ production increased dependently on the increased number of effector cells (Fig. 6). These CTLs recognized the peptide used for stimulation. Namely, the PBMCs stimulated with the SART1$_{690-698}$ produced a higher level of IFN-γ by recognition of this peptide on HLA-A2402-transfected VA13 cells, but not on HLA-A2601-transfected VA13 cells (Fig. 7).

In HLA-A24 heterozygotes, PBMCs from the five of six healthy donors or four of six cancer patients also produced higher levels of IFN-γ by recognition of the KE4 tumor cells as compared to those of either the QG56 or VA13 cells after being stimulated three times with the SART1$_{690-698}$ peptide (see Fig. 5). However, the levels of IFN-γ produced by the PBMCs of the heterozygotes were significantly lower than those produced by the homozygotes in all cases tested. PBMCs from none of the six HLA-A24-/A24-cases ($n = 3$ in healthy donors, $n = 3$ in cancer patients) tested produced a higher level of IFN-γ by

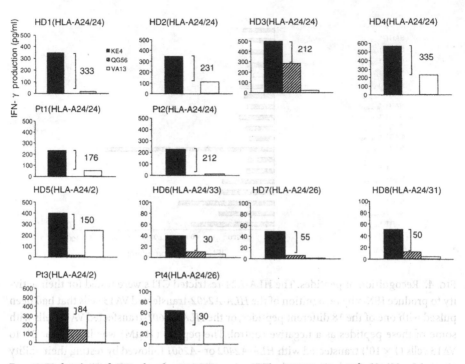

FIG. 5. Induction of CTLs by the SART1$_{690-698}$ peptide. PBMCs (2×10^6) from healthy volunteers (HD 1 to HD 8) and esophageal (Pt 1 to Pt 3) or lung cancer (Pt 4) patients were incubated with the SART1$_{690-698}$ peptide (10 μM) in a well of a 24-well plate containing 2 ml culture medium. At days 7 and 14 of culture, cells were collected, washed, and stimulated with the irradiated autologous PBMCs as APCs that had been preincubated with the same nonapeptide (10 μM) for 2 h. Cells were harvested at day 21 of the culture, and were tested for their activity to produce IFN-γ in response to the KE4 tumor cells (HLA-A2402$^+$, SART1$_{259}$$^+$, SART1$_{800}$$^+$), QG56 (HLA-A24$^-$, SART1$_{259}$$^+$, SART1$_{800}$$^+$), and VA13 (HLA-A24$^-$ and SART1$_{259}$$^-$, SART1$_{800}$$^+$) at an effector-to-target (E : T) cell ratio of 4 : 1 in triplicate assays

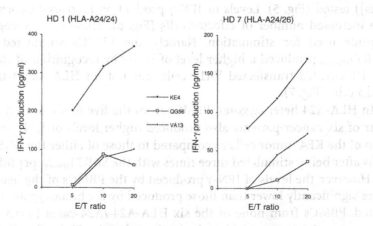

FIG. 6. Dose dependency of peptide-induced CTL activity by PBMCs (*HD 1*, *HD 7*) shown in Fig. 5 was studied at different effector-to-target cell ratios in triplicate assays

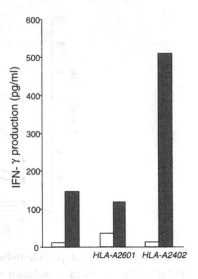

FIG. 7. Recognition of the SART1$_{690-698}$ peptide by CTLs. The PBMCs stimulated by the SART1$_{690-698}$ peptide three times in vitro were tested for their activity to produce IFN-γ by recognition of the *HLA-A2402*-transfected VA13 cells that had been pulsed with the same peptide, or the *HLA-A2601*-transfected VA13 cells with the peptide as a control, at an effector-to-target cell ratio of 4 : 1 in triplicate assays. *White bars*, EYRGFTQDF–; *black bars*, EYRGFTQDF+

recognition of the KE4 tumor cells than by recognition of either the QG56 or VA13 cells (data not shown).

The SART1$_{746-755}$ or SART1$_{758-766}$ peptides were also tested for their activity to induce CTLs in HLA-A24$^+$ donors. However, none of them consistently induced the HLA-A24-restricted CTLs under the conditions employed (data not shown).

These results suggest that the SART1$_{690-698}$ peptide consistently induces the HLA-A24-restricted CTLs recognizing the SART1$_{259}^+$ tumor cells by only three-time stimulation in vitro in PBMCs from both HLA-A24$^+$ cancer patients and healthy donors. CD8$^+$ T cells were then enriched from the PBMCs after the third stimulation with the SART1$_{690-698}$ and were tested for their cytotoxicity to the various target cells in a 6-h ^{51}Cr-release assay (Fig. 8). The CD8$^+$ T cells showed significant levels of lysis only to the KE4 tumor cells, but not to any of the other target cells, including QG56 tumor cells or the autologous Epstein–Barr virus-transformed B-cell line. This study demonstrated that the SART1$_{690-698}$ peptide could induce the HLA-A24-restricted CTLs that recognized the SART1$_{259}^+$ tumor cells in PBMCs of all HLA-A24 homozygotes and the majority of HLA-A24 heterozygotes tested in vitro in both cancer patients and healthy donors.

Four different tumor-rejection antigens (tyrosinase, β-catenin, gp100, and p15) are reported in melanomas to possess antigenic peptides recognized by the HLA-A24-restricted CTLs [24,25]. Some other tumor-rejection antigens identified from melanomas (MAGE-1, MAGE-3, PRAME, and MART-1) might encode HLA-A24-binding antigenic peptides recognized by the CTLs [26–28]. The SART1$_{259}$ antigen was expressed in the majority of esophageal and head and neck SCCs, half of lung adenocarcinomas [13], and one-third of uterine cancers [29]. The HLA-A24 allele is found in approximately 60% of Japanese (the majority, 95%, are genotypically HLA-A2402) [30], 20% of Caucasians, and 12% of Africans [22]. Therefore, the SART1$_{690-698}$ peptide (EYRGFTQDF)

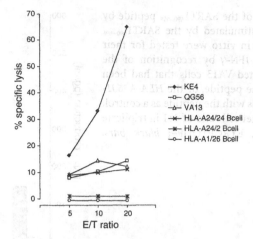

FIG. 8. Cytotoxicity of peptide-induced CD8+ T cells. CD8+ T cells enriched from the PBMCs after the third stimulation with SART1$_{690-698}$ were tested for their cytotoxicity to the various target cells in a 6-h ^{51}Cr-release assay at three different effector-to-target cell ratios. Target cells were the KE4 tumor cells, QG56 lung SCC, and Epstem–Barr virus-(EBV-) transformed B-cell line of the HD 1 (HLA-A24/A24, SART1$_{259}$⁻, and SART1$_{800}$⁺) EBV-transformed B-cell line of the HD 5 (HLA-A24/A2, SART1$_{259}$⁻, and SART1$_{800}$⁺), and EBV-transformed B-cell line of the other HD (HLA-A1/A26, SART1$_{259}$⁻, and SART1$_{800}$⁺), shown by representative results from PBMCs of HD 1 in Fig. 5

could be useful for specific immunotherapy of relatively large numbers of HLA-A24+ cancer patients with SCCs and adenocarcinomas as a cancer vaccine and also as a peptide antigen in vitro to induce CTLs for the adoptive cellular therapy.

Conclusion

The SART1$_{259}$ protein was expressed in the cytosol of the majority of SCCs and one-third to one-half of adenocarcinomas. Because of its preferential expression in the cytosol of SCCs and adenocarcinomas, the SART1$_{259}$ protein, but not SART1$_{800}$, could be a major source of antigenic peptides recognized by the CTLs. The SART1$_{259}$ protein and their peptides could be useful for specific immunotherapy of relatively large numbers of patients with SCCs or adenocarcinomas as cancer vaccine, and also as an antigen in vitro to induce CTLs for adoptive cellular therapy.

Acknowledgments. This study was supported in part by a grant-in-aid from the Ministry of Education, Science, Sports, and Culture of Japan, and a grant from the Ministry of Health and Welfare of Japan.

References

1. van der Bruggen P, Traversari C, Chomez P, Lurquin C, De Plaen E, Van den Eynde B, Knuth A, Boon T (1991) A gene encoding an antigen recognized by cytolytic T lymphocytes on a human melanoma. Science 254:1643–1647
2. Kawakami Y, Eliyahu S, Delgado CH, Robbins PF, Rivoltini L, Topalian SL, Miki T, Rosenberg SA (1994) Cloning of the gene coding for a shared human melanoma antigen recognized by autologous T cells infiltrating into tumor. Proc Natl Acad Sci USA 91:3515–3519
3. Kawakami Y, Eliyahu S, Delgado CH, Robbins PF, Sakaguchi K, Appella E, Yannelli JR, Adema GJ, Miki T, Rosenberg SA (1994) Identification of a human melanoma antigen recognized by tumor-infiltrating lymphocytes associated with *in vivo* tumor rejection. Proc Natl Acad Sci USA 91:6458
4. Brichard V, Van Pel A, Wölfel T, Wölfel C, De Plaen E, Lethé B, Coulie P, Boon T (1993) The tyrosinase gene codes for an antigen recognized by autologous cytolytic T lymphocytes on HLA-A2 melanomas. J Exp Med 178:489–495
5. Traversari C, van der Bruggen P, Luescher IF, Lurquin C, Chomez P, Van Pel A, De Plaen E, Amar-Costesec A, Boon T (1992) A nonapeptide encoded by human gene MAGE-1 is recognized on HLA-A1 by cytolytic T lymphocytes directed against tumor antigen MZ2-E. J Exp Med 176:1453–1457
6. Cox AL, Skipper J, Chen Y, Henderson RA, Darrow TL, Shabanowitz J, Engelhard VH, Hunt DF, Slingluff CL Jr (1994) Identification of a peptide recognized by five melanoma-specific human cytotoxic T cell lines. Science 264:716–719
7. Kawakami Y, Eliyahu S, Sakaguchi K, Robbins PF, Rivoltini L, Yannelli JR, Appella E, Rosenberg SA (1994) Identification of the immunodominant peptides of the MART-1 human melanoma antigen recognized by the majority of HLA-A2-restricted tumor infiltrating lymphocytes. J Exp Med 180:347–352
8. Marchand M, Weynants P, Rankin E, Arienti F, Belli F, Parmiani G, Cascinelli N, Bourlond A, Vanwijck R, Humblet Y, Canon J-L, Laurent C, Naeyaert J-M, Plagne R, Deraemaeker R, Knuth A, Jager E, Brasseur F, Herman J, Coulie PG, and Boon T (1995) Tumor regression responses in melanoma patients treated with a peptide encoded by gene MAGE-3. Int J Cancer 63:883–888
9. Rosenberg SA, Yang JC, Schwartzentruber DJ, Hwu P, Marincola FM, Topalian SL, Restifo NP, Dudley ME, Schwarz SL, Spiess CA, Einhorn JH, White DE (1998). Immunologic and therapeutic evaluation of a synthetic peptide vaccine for the treatment of patients with metastatic melanoma. Nat Med 4:321–327
10. Nestle FO, Alijagic S, Gilliet M, Sun Y, Grabbe S, Dummer R, Burg G, Schadendorf D (1998) Vaccination of melanoma patients with peptide- or tumor lysate-pulsed dendritic cells. Nat Med 4:328–332
11. Mandruzzato A, Brasseur F, Andry G, Boon T, van der Bruggen P (1997) A CASP-8 mutation recognized by cytotoxic T lymphocytes on a human head and neck carcinoma. J Exp Med 186:785–793
12. Jäger E, Chen Y-T, Drijfhout JW, Karbach J, Ringhoffer M, Jäger D, Arand M, Wada H, Knuth A (1998) Simultaneous humoral and cellular immune response against cancer-testis antigen NY-ESO-1: definition of human histocompatibility leukocyte antigen (HLA)-A2-binding peptide epitopes. J Exp Med 187:265–270
13. Shichijo S, Nakao M, Imai Y, Takasu H, Kawamoto M, Niiya F, Yang D, Toh Y, Yamana H, Itoh KA (1998) Gene encoding antigenic peptides of human squamous cell carcinoma recognized by cytotoxic T lymphocytes. J Exp Med 187:277–288

14. Kikuchi K, Nakao M, Inoue Y, Matsunaga K, Shichijyo S, Itoh K (1999) Identification of a SART-1-derived peptide capable of inducing the HLA-A24-restricted and tumor-specific cytotoxic T lymphocytes. Int J Cancer 81:459–466
15. Nakao M, Yamana H, Imai Y, Toh Y, Toh U, Kimura A, Yanoma S, Kakegawa T, Itoh K (1995) Cancer Res 55:4248–4252
16. Kozak M (1989) The scanning model for translation: an update. J Cell Biol 108: 229–241
17. Larsen B, Peden J, Matsufuji S, Matsufuji T, Brady K, Maldonado R, Wills NM, Fayet O, Atkins JF, Gesteland RF (1995) Upstream stimulators for recoding. Biochem Cell Biol 73:1123–1129
18. Pilipenko EV, Gmyl AP, Maslova SV, Svitkin YV, Sinyakov AN, Agol VI (1992) Prokaryotic-like cis elements in the cap-independent internal initiation of translation on picornavirus RNA. Cell 68:119–131
19. Wang R-F, Parkhurst MR, Kawakami Y, Robbins PF, Rosenberg SA (1996) Utilization of an alternative open reading frame of a normal gene in generating a novel human cancer antigen. J Exp Med 183:1131–1140
20. Ossipow V, Descombes P, Schibler U (1993) CCAAT/enhancer-binding protein mRNA is translated into multiple proteins with different transcription activation potentials. Proc Natl Acad Sci USA 90:8219–8223
21. Goulder P, Conlon C, McIntyre K, McMichael A (1994) Identification of a novel human leukocyte antigen A26-restricted epitope in a conserved region of Gag. AIDS 10:-1441–1443
22. Imanishi T, Akaza T, Kimura A, Tokunaga K, Gojobori T (1992) Allele and haplotype frequencies for HLA and complement loci in various ethnic groups. In: Tsuji K, Aizawa M, Sasazuki T, (eds) HLA 1991, vol 1. Oxford Scientific, Oxford, pp 1065–1220
23. Date Y, Kimura A, Kato H, Sasazuki T (1996) DNA typing of the HLA-A gene: population study and identification of four new alleles in Japanese. Tissue Antigens 47:93–101
24. Kang X, Kawakami Y, El-Gamil M, Wang R, Sakaguchi K, Yanneli E, Appella E, Rosenberg SA, Robbins PF (1995) Identification of a tyrosinase epitope recognized by HLA-A24-restricted, tumor-infiltrating lymphocytes. J Immunol 155:1343–1348
25. Robbins PF, El-Gamil M, Li YF, Topalian SL, Rivoltini L, Sakaguchi K, Appella E, Rosenberg SA (1995) Cloning of a new gene encoding an antigen recognized by melanoma-specific HLA-A24-restricted tumor-infiltrating lymphocytes. J Immunol 154:5944–5950
26. Celis E, Fikes J, Wentworth P, Sidney J, Southwood S, Maewal A, Guercio MFD, Sette A, Livington B (1994) Identification of potential CTL epitopes of tumor-associated antigen MAGE-1 for five common HLA-A alleles. Mol Immunol 31:1423–1430
27. Ikeda H, Lethe B, Lehmann F, Baren NV, Baurain JF, Smet CD, Chambost H, Vitale M, Moretta A, Boon T, Coulie PG (1997) Characterization of an antigen that is recognized on a melanoma showing partial HLA loss by CTL expressing an NK inhibitory receptor. Immunity 6:199–208
28. Tanaka F, Fujie T, Tahara K, Mori M, Takesako K, Sette A, Celis E, Akiyoshi T (1997) Induction of antitumor cytotoxic T lymphocytes with a MAGE-3-encoded synthetic peptide presented by human leukocytes antigen-A24. Cancer Res 57:4465–4468
29. Matsumoto M, Shichijo S, Kawano S, Nishida T, Sakamoto M, Itoh K (1998) Expression of the SART-1 antigens in uteric cancers. J Cancer Res 89:1292–1295
30. Tokunaga K, Ishikawa Y, Ogawa A, Wang H, Matsunaga S, Morinaga S, Lin L, Bannai M, Watanabe Y, Kashiwase K, Tanaka H, Akaza T, Tadokoro K, Juji T (1997) Sequence-based association analysis of HLA class I and II alleles in Japanese supports conservation of common haplotypes. Immunogenetics 46:199–205

Adoptive Immunotherapy of Human Diseases with Antigen-Specific T-Cell Clones

Stanley R. Riddell[1,2], Edus H. Warren[1], Deborah Lewinsohn[1], Scott Brodie[2], Rici de Fries[1], Lawrence Corey[1,2], and Philip D. Greenberg[1,2]

Summary. Cellular adoptive immunotherapy refers to the transfer of effector cells of the immune system to treat infectious or malignant diseases. Our laboratory has been investigating the use of adoptive immunotherapy with antigen-specific T-cell clones to prevent cytomegalovirus (CMV) infection in allogeneic bone marrow transplant (BMT) recipients; augment immune responses to human immunodeficiency virus (HIV) in HIV-seropositive individuals; and to induce a graft-versus-leukemia (GVL) response in allogeneic BMT recipients. In each of these settings the rationale for developing adoptive immunotherapy is based on clear evidence that deficiencies of antigen-specific T cells are responsible for disease progression. Functional studies of effector activity and genetic markers have been used to demonstrate that adoptively transferred T-cell clones can persist and function in the host and migrate appropriately to sites of disease. It is anticipated that additional studies will define disease settings in which T-cell therapy can be beneficial and further elucidate the requirements for effective immunotherapy in humans.

Key words. Adoptive immunotherapy, Cytomegalovirus, Human immunodeficiency virus, Graft-versus-leukemia, Minor histocompatibility antigens

Introduction

Our laboratory has been investigating adoptive immunotherapy with antigen-specific T-cell clones in clinical settings in which deficiencies of T-cell immunity are responsible for disease progression. The initial studies have focused

---

[1] Fred Hutchinson Cancer Research Center, 1100 Fairview Avenue North, Seattle, WA 98109, USA
[2] University of Washington, Department of Medicine, Seattle, WA 98104, USA

on developing and evaluating T-cell immunotherapy for viral diseases including cytomegalovirus (CMV) infections that occur in allogeneic bone marrow transplant (BMT) recipients and progressive human immunodeficiency virus (HIV) infection. This approach is also being developed as a strategy to prevent or treat leukemic relapse after allogeneic BMT using T-cell clones specific for minor histocompatibility antigens to augment the graft-versus-leukemia (GVL) effect. As discussed in this review, the clinical application of T-cell therapy in these settings is still in its early stages and efficacy has not yet been established. However, these studies have provided insights into the requirements for effectively augmenting T-cell immunity in humans by adoptive immunotherapy and serve as a basis for continued development of this approach for viral and malignant diseases.

Adoptive T-Cell Therapy of Cytomegalovirus Infection in Allogeneic BMT Recipients

Cytomegalovirus is acquired by the majority of the population in childhood and typically causes an uneventful primary infection. However, the virus persists in the host in a state of latency that is maintained by CD4+ and CD8+ T cells reactive with CMV antigens. Allogeneic BMT recipients who receive doses of chemotherapy or radiotherapy sufficient to ablate the immune system before transplantation, and immunosuppressive drugs to prevent graft-versus-host disease (GVHD) post transplant, have a prolonged deficiency of T-cell immunity. Reactivation of CMV replication occurs frequently in these patients and may progress to CMV interstitial pneumonia (CMV-IP) if antiviral therapy is not instituted. The peak incidence of CMV-IP is at 2 months after transplant, and more than 50% of the patients have a fatal outcome [1]. Analysis of the tempo of recovery of CD4+ and CD8+ CMV-specific T-cell responses in BMT recipients demonstrated a strong correlation between the absence of CD8+ cytotoxic T cells (CTL) reactive with CMV antigens and the progression of CMV infection to CMV-IP [2,3]. This result suggested that restoring a competent T-cell response to CMV by the adoptive transfer of CMV-specific T-cell clones isolated from the immunocompetent bone marrow donor and exanded by in vitro culture might be beneficial for containing CMV infection.

Clinical Protocol for Adoptive Transfer of CD8+ CMV-Specific T-Cell Clones

A phase 1 study was designed to evaluate the safety and immunomodulatory effects of administering donor-derived CD8+ CMV-specific CTL clones and enrolled 14 patients who had received an allogeneic BMT for a CMV seropositive donor [4,5]. These patients were treated with four weekly intravenous infusions of T-cell clones in escalating doses of $3.3 \times 10^7/m^2$, $1 \times 10^8\,m^2$, $3.3 \times 10^8/m^2$, and $1 \times 10^9/m^2$ of body surface area, respectively beginning 28–42 days after BMT. No serious acute or late toxicities were attributed to the CTL infusion, and the T cells were routinely administered in the Outpatient Department.

Transfer of CD8+ CMV-Specific CTL Provides Durable Reconstitution of CMV Immunity

Peripheral blood samples were analyzed at multiple time points before, during, and after the CTL infusions to determine if immune responses were restored and the duration that transferred CTL persisted in vivo. Three patients had evidence of endogenous recovery of CMV-specific CTL responses at the time adoptive immunotherapy was initiated, and in these patients it was difficult to accurately define the contribution of transferred CTL to host immunity. However, the remaining 11 patients treated on the study were completely deficient in CMV-specific CTL immediately before the first infusion. Each of these patients exhibited a detectable response 2 days after the first infusion, and this response was augmented with each subsequent infusion. By the end of the 4-week treatment period, levels of CTL activity equivalent to those in the donor were achieved (Fig. 1).

Analysis of the durability of CTL responses in the recipients of adoptive immunotherapy revealed that all patients maintained responses for at least 3 months after therapy. However, the magnitude of the CTL response declined in the subset of patients who had developed GVHD as a consequence of the allogeneic BMT and required prolonged intensive immunosuppression with both cyclosporine and prednisone. Although this may reflect the effect of prednisone on the viability of transferred CTL, these patients also remained deficient in endogenous CD4+ Th responses to CMV. In animal models, the persistence and function of virus-specific CD8+ CTL has been shown to require CD4+ T cells [6,7], suggesting the persistence of transferred CD8+ CTL may have been compromised by the absence of CD4+ CMV-specific Th.

The phase 1 trial provided critical insights into the potential to use virus-specific T-cell clones to restore or augment T-cell immunity in humans. None of the 14 patients on the study developed CMV disease [5], but the trial was not of sufficient size for definitive analysis of the protective efficacy of this approach. To address this issue, we have now initiated a larger phase 2 trial in which patients are receiving both CMV-specific CD8+ CTL and CD4+ Th clones. Preliminary results demonstrate that CD4+ Th clones can also be safely transferred, restore CMV-specific Th immunity, and in patients not

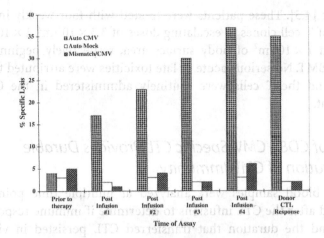

FIG. 1. Cell dose-related reconstitution of CD8+ CMV-specific cytotoxic T-cell (CTL) responses after adoptive immunotherapy. Peripheral blood lymphocytes were obtained from patients before and 2 days after each infusion of cytomegalovirus (CMV) -specific CD8+ CTL clones, and CMV-specific CTL generation was assessed as previously described [5]. Target cells include autologous CMV-infected, autologous mock-infected, and class I MHC-mismatched CMV-infected fibroblasts. Data are shown for an effector-to-target ratio of 10:1

receiving prednisone, provide effective helper function for transferrred CTL (S.R. Riddell and P.D. Greenberg, unpublished data).

Adoptive Immunotherapy of HIV Infection

Several lines of evidence suggests that CD8+ CTL, which are capable of elim- inating productively infected cells by direct cytolysis and of limiting viral spread by secretion of chemokines, contribute to the host control of HIV replication. After primary infection, the appearance of HIV-specific CTL coincides with a reduction in viremia and return to near-normal CD4 counts [8–10]. In chronically infected individuals, plasma viral load is inversely correlated with the magnitude of the HIV-specific CD8+ CTL response [10,11]. The subset of infected individuals, termed long-term nonprogressors, who remain asymptomatic and retain normal CD4 counts for several years exhibit strong and persistent CD8+ Gag-specific CTL responses [12] and have virus isolates that remain sensitive to CTL [13]. However, virus replication is not completely controlled by the host CTL response, and a progressive CD4+ T-cell immunodeficiency characterized by high levels of HIV and a complete absence of circulating HIV-specific CTL eventually ensues [8,14].

Clinical Protocol for Adoptive Transfer of CD8+ HIV-Specific CTL Clones

Based on these findings that CD8+ HIV-specific CTL mediate considerable antiviral activity and become depleted with progressive infection, we have evaluated whether the adoptive transfer of large numbers of autologous CD8+ HIV-specific CTL clones isolated and expanded in vitro could augment the partially effective host response and exert additional antiviral activity. HIV Gag-specific CD8+ CTL clones were isolated from six HIV seropositive individuals on a stable regimen of antiretroviral drug therapy, with CD4+ cell counts between 200 and 500 cells/mm^3 and without evidence of opportunistic infection. To facilitate monitoring of cell persistence and localization to sites of infection in vivo, the three infusions of unmodified CTL were followed by two infusions of CTL clones that had been transduced with a retrovirus (LN) containing the neomycin phosphotransferase (*neo*) gene. The CTL clones were MHC class I restricted and recognized epitopes in either the p24 or p17 subunit of HIV Gag (Table 1). Three infusions of unmodified CTL clones were administered at 2-week intervals in doses of 3.3×10^8 cells/m^2, 1×10^9 cells/m^2, and 3.3×10^9 cells/m^2. One week later, two infusions of LN-modified CTL were given at a 1-week interval in doses of 1×10^9 cells/m^2 and 3.3×10^9 cells/m^2, and in a subset of three patients, a lymph node biopsy was performed 4 days after the final infusion.

CD8+ CTL produce inflammatory cytokines following stimulation with antigen, and acutely augmenting the host response by transferring large numbers of virus-specific CTL to patients with a high viral load might cause toxicity. Indeed, all patients with a measurable viral load had fever, myalgias, or arthralgias following some of the infusions. However, the symptoms lasted less than 48h and only required symptomatic treatment. No significant changes in CD4+ T-cell count were noted during, or in the 3 months following, CTL immunotherapy.

Adoptive Transfer of CD8+ HIV-Specific CTL Transiently Provides High Levels of HIV-Specific Cytolytic Activity

Peripheral blood mononuclear cells (PBMC) were obtained 1 day before and 1, 6, and 14 days after infusion of the highest T-cell dose (3.3×10^9 cells) and assayed for lytic activity against autologous Epstein–Barr virus-(EBV-) transformed lymphoblastoid cell lines (LCL) expressing either HIV Gag or an irrelevant CMV protein. HIV-specific cytolytic activity was not detected by direct assay of the pretherapy PBMC. However, at 1 and 6 days after the T-cell infusion, high levels of HIV-specific cytolytic activity were detected in PBMC. The increased lytic activity returned to pre-therapy baseline levels by day 14 (Fig. 2).

TABLE 1. Epitope specificities of HIV-specific CD8+ cytotoxic T-cell (CTL) clones used in adoptive immunotherapy

Patient	Clone	MHC restriction	Gag protein	Peptide epitope
1	NT-10D10*	B44	p24	AEQASQEVKNW (aa 306–316)
	NT-16A7*			
	NT-59D9*			
	LN-2A3*			
	LN-21D12*			
	LN-32E4*			
2	NT-9A6*	B51	p15	
	NT-40A8*	B44	p24	
	LN-6E11	B51	p15	
3	NT-44H8*		p24	NAWVKVVEEKAFSPEVIPMF (aa 153–172)
	NT-34D12		p24	VHQAISPRTLNAWVKVVEEK (aa 142–162)
	LN-33D7*	B57		
	LN-34H10*		p24	NAWVKVVEEKAFSPEVIPMF (aa 153–172)
	LN-36C10*			
	LN-41A8		p24	GSDIAGTTSTLQEQIGWMTN
4	NT-30F5*	A3	p17	KIRLRPGGK (aa 18–26)
	NT-52H8*	A3	p17	RLRPGGKKK (aa 20–28)
	NT-8A3	B8	p24	EIYKRWII (aa 260–267)
	NT-31D11	B8	p24	GHQAAMQMLKETINEEAAEW (aa 193–212)
	LN-9E12*	A3	p17	KIRLRPGGK (aa 18–26)
	LN16B11	B8	p24	EIYKRWII (aa 260–267)
5	NT-35D1*			
	NT-19H9*	A2	p17	GSEELRSLYNTVATL (aa 71–85)
	LN-48D1*	B7	p24	SPRTLNAWV (aa 148–156)
6	NT-3G7*	B62	p17	RLRPGGKKKY (aa 20–29)
	LN-18H3*			

The frequency of *neo*-modified CD8+ T cells in the peripheral blood samples obtained after CTL transfer was determined by flow cytometry using a novel in situ hybridization (ISH) assay to detect *neo* sequences in fixed cells subjected to PCR amplification with *neo* amplimers [15]. One day following infusion of the highest dose of *neo*-modified CTL (3.3×10^9 cells/m^2), *neo*+ T cells comprised from 2% to 3.5% of CD8+ T cells in the peripheral blood. Consistent with the analysis of Gag-specific cytolytic activity in the PBMC after T-cell transfer, the frequency of *neo*-modified CD8+ T cells declined each day after the infusion and were no longer detected by flow cytometry at 2–3 weeks following the final T-cell infusion [15]. The short in vivo persistence of transferred CTL may reflect the absence of CD4+ HIV-specific Th, and studies

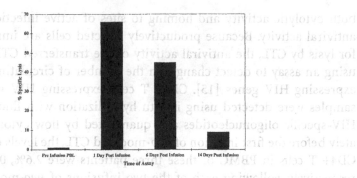

FIG. 2. Adoptive transfer of CD8+ HIV-specific CTL clones augments Gag-specific cytolytic activity in the peripheral blood (representative patient). Peripheral blood lymphocytes were obtained from the patient before the infusion of the highest cell dose ($3.3 \times 10^9/m^2$) and 1, 6, and 14 days after the T-cell infusion. The lymphocytes were assayed directly at an effector-to-target ratio of 100:1 for the ability to lyse chromium-labeled autologous Epstein–Barr Virus- (EBV-) transformed lymphocytes (LCL) infected with either vaccinia/Gag recombinant virus or a control vaccinia virus. The Gag-specific cytolytic activity is calculated as the difference in lysis between vac/Gag and vac targets

are now in progress to determine if T-cell survival can be improved with the addition of IL-2.

Transferred HIV-Specific CTL Migrate to Lymph Nodes and Colocalize at Sites of HIV Replication

Migration of the transferred CTL to lymph nodes and their relationship with HIV-infected cells within the lymph node were examined in serial tissue sections made from lymph node biopsies obtained 4 days after the last infusion of LN-modified CTL. In situ PCR with *neo* primers was used to amplify *neo* sequences in single cells followed by hybridization with a digoxigenin-labeled, *neo*-specific probe to localize the *neo'* CTL. Direct hybridization with digoxigenin-labeled HIV-specific riboprobes for HIV RNA was used to local-ize HIV-infected cells [15]. This latter technique does not detect latently infected cells because more than 20 copies of viral RNA per cell are required for a positive signal; thus, only cells HIV actively replicating HIV are detected. These studies demonstrated that adoptively transferred HIV-specific CTL and cells productively infected with HIV colocalized to discrete areas in the parafollicular regions of the lymph node [15].

Antiviral Activity of Adoptively Transferred HIV-Specific CD8+ CTL

Three of the six patients enrolled in this study had measurable levels of HIV before therapy. It was anticipated the transferred T cells, which demonstrated both cytolytic activity and homing to sites of active infection, would have antiviral activity. Because productively infected cells are immediate targets for lysis by CTL, the antiviral activity of the transferred CTL was analyzed using an assay to detect changes in the number of circulating CD4+ T cells expressing HIV genes [15]. CD4+ T cells expressing HIV genes in PBMC samples were detected using in situ hybridization with fluorescein-labeled HIV-specific oligonucleotides and quantitated by flow cytometry. Immediately before the first infusion of *neo*-modified CTL, the levels of HIV-infected CD4+ T cells in PBMC of these three patients were 2.6%, 0.5%, and 2.4%, respectively. Following each of the two infusions of *neo*-modified CTL, the percentage of productively infected CD4+ T cells declined dramatically over the ensuing 3 days. A similar decline in productively infected CD4+ T cells was observed following the infusions of unmodified CTL. However, the frequency of transferred CTL decreased over time, and the percentage of productively infected CD4+ T cells rebounded to baseline levels by 7 days after each infusion. These results provide direct evidence for the antiviral effects of CD8+ CTL and suggest that improving T-cell persistence may result in a sustained antiviral effect.

Plasma HIV, as measured by branched-chain DNA (bDNA) assay, did not significantly change during immunotherapy apart from a transient increase noted in the first 24 h after therapy. This result potentially reflected release of preformed virions from target cells lysed by CTL or increased viral production by neighboring infected cells caused by proinflammatory cytokines released by the CTL. The persistence of transferred CTL in these patients was short lived, and replenishment of the infected pool of CD4+ T cells may have occurred before reductions in plasma virus could be observed. The production of virus by cells harboring viral variants containing mutations in sequences encoding CTL epitopes could also potentially contribute to the sustained level of plasma virus. However, analysis of the sequence of the predominant plasma viral isolates, which were sensitive to the CTL administered at the start of therapy, did not reveal the accumulation of escape mutations during and following T-cell therapy (D.A. Lewinsohn and S.R. Riddell, unpublished data).

Minor Histocompatibility Antigens as Targets for T-Cell Immunotherapy to Augment Graft-Versus-Leukemia Reactions After Allogeneic Bone Marrow Transplant

GVHD occurs in 30%–40% of allogeneic BMT recipients even when the donor and recipient are identical at the MHC because of differences at minor H loci [16]. In addition to being targets for GVHD, minor H antigens are presumed to be the targets for the GVL effect observed after allogeneic BMT. Thus, removal of T cells from the donor marrow reduces the incidence of GVHD but results in increased leukemic relapse [17]. By contrast, the infusion of donor lymphocytes to patients who have relapsed following allogeneic BMT will induce a complete remission in the majority of patients with chronic myeloid leukemia (CML) and a significant fraction of patients with acute myeloid leukemia (AML) and acute lymphocytic leukemia (ALL) [18]. Unfortunately, the infusion of unselected donor lymphocytes is usually complicated by GVHD. A crucial issue is to determine if T-cell therapy to induce GVL responses could be devised to target minor H antigens that are not expressed on tissues which are targets of GVHD.

T-cell clones reactive with human minor H antigens and potentially involved in GVHD, GVL, or rejection responses have been isolated from patients receiving MHC-matched BMT, solid organ transplants, or multiple blood transfusions. Recent studies assessing recognition of cells derived from different tissues by such minor H antigen-specific CTL suggest that some minor H antigens are expressed on only a limited range of host tissues, presumably reflecting tissue-specific functions of the proteins [19–21]. Thus, minor H antigens that are only expressed by hematopoietic cells including leukemic progenitors could potentially be targets for adoptive immunotherapy to promote a GVL response without causing GVHD.

Human Minor H Antigens Defined by T-Cell Clones

Recent studies suggest that the number of minor H loci in humans is likely to be very large and greatly exceed the diversity of the MHC. Minor H antigens can be broadly categorized into those encoded by autosomal genes and those encoded by the Y chromosome. Although only relevant in the setting of sex-mismatched BMT, several T-cell clones that recognize minor H antigens encoded by the Y chromosome (H-Y antigens) and presented by either HLA A1, A2, B7, or B8 have been isolated and characterized [22]. The human SMCY gene was shown to encode the peptides recognized by the HLA A2- and HLA B7-restricted H-Y-specific T-cell clones isolated by Goulmy et al. [23,24]. SMCY is ubiquitously expressed and cells of both hematopoietic and nonhematopoietic derivation are recognized by T cells specific for SMCY-encoded antigens. Thus, SMCY is a potentially important target for GVHD but is unlikely to be a suitable target for T-cell therapy to induce a selective GVL response.

CD8+ T-cell clones specific for a novel H-Y antigen have been isolated by our group after transplantation of marrow from a female donor into a male recipient. These H-Y-specific CTL were restricted by HLA B8 and lysed hematopoietic cells including leukemic blasts obtained from male HLA B8+

donors, but not skin fibroblasts from the same donors, suggesting this antigen may be a suitable target for GVL therapy [19]. Mapping studies to identify the Y chromosome gene encoding this antigen have been performed in collaboration with Elizabeth Simpson and David Page. Cell lines derived from males with deletions of portions of the Y chromosome were used as target cells to demonstrate that the gene encoding the HLA B8-restricted H-Y antigen was distinct from SMCY and was located on a region of Y known to encode five genes (E.H. Warren and S.R. Riddell, unpublished data). Transfection studies to identify the gene, now in progress, will permit molecular characterization of the tissue expression of this novel H-Y antigen and determine if it is truly limited to hematopoietic cells.

Several groups have isolated CD8+ CTL specific for minor H antigens encoded by autosomal genes, and initial studies of the frequency of the alleles encoding these antigens in the population and their tissue expression have identified several candidates as targets for GVL therapy. Goulmy et al. have characterized CD8+ CTL clones that define seven minor H antigens, designated HA-1 to HA-7, encoded by autosomal genes [22]. The HA-3, -4, -5, -6, and -7 antigens appear by in vitro cytotoxicity studies to be ubiquitously expressed and not suitable targets for inducing a selective GVL effect [21]. HA-1 and HA-2 were not expressed in nonhematopoietic cells such as keratinocytes, fibroblasts, and renal epithelial cells. However, one study has suggested that mismatching of HA-1 correlates with the development of grade 2–4 GVHD, raising some doubt as to its potential as a target to induce a selective GVL response [25]. This finding highlights the importance of gene identification and molecular analysis of tissue expression in vivo.

A study in our lab involving ten allogeneic BMT donor-recipient pairs isolated and characterized CD8+ CTL clones specific for 19 minor H antigens encoded by autosomal genes [19]. Ten of these 19 minor H antigens were presented by HLA A2 or B7 but are distinct from HA-1, -2, -4, -5, -6, and -7 on the basis of tissue expression or population frequency. The other 9 minor H antigens were presented by class I HLA molecules not previously described as restricting alleles for minor H antigen-specific T cells, including HLA A3, A11, A29, B44, B53, and Cw7 [19]. Five of these 19 minor H antigens are expressed in both hematopoietic cells and skin fibroblasts while the other 14 are only expressed in hematopoietic cells including AML, ALL, and CLL samples. Efforts at gene identification are now under way to further define the utility of these minor H antigens as targets for GVL therapy.

Expression of Minor H Antigens on Leukemic Cells

Studies assessing recognition of fresh leukemic cells by minor H antigen-specific T-cell clones have demonstrated that such CTL will lyse leukemic blasts in vitro and inhibit the outgrowth of leukemic colonies in soft agar

[26,27]. While these results demonstrate that minor H antigens are expressed on a proportion of the leukemic blast population, they do not define the susceptibility of the leukemic stem cell to recognition by CTL. By inoculating AML samples into immunodeficient mice, a putative leukemic stem cell termed the severe combined immuno deficiency (SCID) leukemia-initiating cell (SL-IC), which resides in the CD34+ CD38− fraction and is present in very low frequency in blood or bone marrow from AML patients has been identified by Dick et al. [28]. The rarity of SL-IC makes it difficult to assess the susceptibility of this cell to lysis by minor H antigen-specific T cells using in vitro assays. Thus, in collaboration with John Dick, we utilized the NOD/SCID mouse transplant model to determine if the AML stem cell is a target for minor H antigen-specific CTL. These studies showed that the engraftment of AML was prevented by CD8+ CTL clones specific for minor H antigens, demonstrating that SL-IC can be eliminated by CTL (D. Bonnet and E.H. Warren, manuscript submitted). Engraftment of stem cells that do not express the minor H antigen was not affected by CTL, suggesting that administration of minor H antigen-specific T cells to BMT recipients would not impair engraftment of donor hematopoietic cells.

Identification of Minor H Genes and Peptides

Identification of the genes encoding minor H antigens will greatly facilitate the selection of reactive T cells for use in augmenting GVL responses. Three strategies are being employed to identify the genes encoding minor H antigens. The first strategy, which has been successfully applied to the identification of minor H antigens in mice, utilizes positional cloning techniques to define the chromosome location of the gene [29,30]. One study in humans used a panel of cells obtained from two large families that had previously been mapped extensively for polymorphic loci to localize the gene encoding an HLA B7-restricted minor H antigen to chromosome 22 [31]. However, considerable effort is still necessary to identify the specific gene and to derive the sequence of the antigenic epitope.

The second approach to gene identification begins by obtaining the sequence of the peptide epitope. Peptides are eluted from cell-surface MHC molecules, separated using biochemical techniques to identify the fraction containing the minor H antigen peptide, and sequenced using mass spectrometry [32]. The nucleotide sequence encoding the epitope can be deduced from the amino acid sequence and databases containing human DNA sequences searched to determine if the epitope is encoded by a known gene. This method has been used to identify the amino acid sequence of four human minor H antigens, two H-Y antigens presented by HLA A2 and HLA B7, respectively, and the HA-1 and HA-2 antigens encoded by autosomal genes and presented by HLA A2 [23,33,34].

The third approach to identify genes encoding CD8+ CTL-defined minor H antigens, which is being employed in our lab and others, uses a strategy pioneered by Boon et al. for identifying genes encoding tumor antigens [35]. This technique involves cotransfection of antigen-negative target cells with pools of cDNA from a library derived from an antigen-positive cell and a cDNA encoding the class I MHC-restricting allele. The proteins expressed by the introduced cDNAs are processed and the derived peptides presented with the introduced class I MHC molecule. Recognition of the transfected cells by T cells can be assessed by cytokine release into the media, and cDNA pools inducing a positive response are then progressively subdivided and screened in an identical fashion until a single cDNA encoding the antigen is identified. This cDNA expression cloning strategy, which is complementary to the peptide elution approach and may be more broadly applicable, has recently been used to identify the gene encoding HB-1 [36] and the gene encoding an HLA A3-restricted minor H antigen identified by our group (M. Gavin, E.H. Warren, and S.R. Riddell, unpublished data).

Development of a Clinical Protocol to Evaluate Adoptive Immunotherapy with Minor H Antigen-Specific T-Cell Clones to Induce GVL Activity

Relapse of acute leukemia remains a major obstacle to successful allogeneic BMT, and survival is not improved by increasing the intensive chemoradiotherapy preparative regimen because of increased toxicity [37]. The high relapse rate in patients with advanced AML or ALL makes it feasible to design T-cell therapy trials in which T-cell clones specific for recipient minor H antigens are prospectively isolated for use when relapse occurs. Clones with reactivity for recipient hematopoietic cells including leukemic blasts but not for nonhematopoietic cells would be selected and cryopreserved. If the patient relapses, the clones could be thawed and expanded for infusion. To avoid causing GVHD, it would be preferable to only use T-cell clones specific for minor H antigens encoded by genes known to be selectively expressed in hematopoietic cells, but until additional genes encoding minor H antigens are identified, a complete molecular analysis is not always available.

Developments in gene therapy have provided novel suicide genes that could be used to improve the safety of therapy with T-cell clones specific for minor H antigens that have not yet been characterized at the molecular level. For example, the introduction of the HSV-TK gene into T-cell clones permits their ablation in vivo by the administration of ganciclovir. This strategy has already been used successfully to ameliorate GVHD occurring following adoptive immunotherapy with unselected polyclonal donor lymphocytes [38]. A problem with the expression of HSV-TK in transferred T cells is the potential for the premature elimination of transferred cells by an immune response elicited to the TK transgene product [39], although repeated infusions of unmodified T cells could be administered if the initial infusions of transduced cells were not associated with GVHD and mediated GVL activity.

Conclusions

The reconstitution of CD8+ and CD4+ CMV-specific T-cell responses in imunodeficient allogeneic BMT recipients by adoptive immunotherapy highlights the potential of this approach for augmenting immune responses in humans. The demonstrations that adoptively transferred HIV-specific T-cell clones home to sites of infection in lymph nodes suggests that prolonged ex vivo culture does not prevent T cells from migrating appropriately in vivo. Future studies will apply T-cell therapy for human malignancies. One of the most attractive opportunities is the use of T-cell clones specific for minor H antigens to prevent or treat leukemic relapse after allogeneic BMT. It is anticipated that future investigation of this approach will continue to define the requirements to successfully utilize specific T-cell therapy in humans.

References

1. Meyers JD, Ljungman P, Fisher LD (1990) Cytomegalovirus excretion as a predictor of cytomegalovirus disease after marrow transplantation: importance of cytomegalovirus viremia. J Infect Dis 162:373–380
2. Reusser P, Riddell SR, Meyers JD, Greenberg PD (1991) Cytotoxic T-lymphocyte response to cytomegalovirus after human allogeneic bone marrow transplantation: pattern of recovery and correlation with cytomegalovirus infection and disease. Blood 78:1373–1380
3. Li CR, Gilbert MJ (1994) Recovery of HLA-restricted cytomegalovirus (CMV)-specific T-cell responses after allogeneic bone marrow transplant: correlation with CMV disease and effect of ganciclovir prophylaxis. Blood 83:1971–1979
4. Riddell SR, Watanabe KS, Goodrich JM, Li CR, Agha ME, Greenberg PD (1992) Restoration of viral immunity in immunodeficient humans by the adoptive transfer of T cell clones. Science 257:238–241
5. Walter EA, Greenberg PD, Gilbert MJ, Finch RJ, Watanabe KS, Thomas ED, Riddell SR (1995) Restoration of cellular immunity against CMV in recipients of allogeneic bone marrow by transfer of T-cell clones from the donor. N Engl J Med 333:1038–1044
6. Matloubian M, Concepcion RJ, Ahmed R (1994) CD4+ T cells are required to sustain CD8+ cytotoxic T-cell responses during chronic viral infection. J Virol 68:8056–8063
7. Zajac AJ, Blattman JN, Murali-Krishna K, Sourdive DJ, Suresh M, Altman JD, Ahmed R (1998) Viral immune evasion due to persistence of activated T cells without effector function. J Exp Med 188:2205–2213
8. Koup RA, Safrit JT, Cao Y, Andrews CA, McLeod G, Borkowsky W, Farthing C, Ho DD (1994) Temporal association of cellular immune responses with the initial control of

viremia in primary human immunodeficiency virus type 1 syndrome. J Virol 68:4650–4655

9. Borrow P, Lewicki H, Hahn BH, Shaw GM, Oldstone MB (1994) Virus-specific CD8+ cytotoxic T-lymphocyte activity associated with control of viremia in primary human immunodeficiency virus type 1 infection. J Virol 68:6103–6110

10. Musey L, Hughes J, Schacker T, Shea T, Corey L, McElrath MJ (1997) Cytotoxic-T-cell responses, viral load, and disease progression in early human immunodeficiency virus type 1 infection. N Engl J Med 337:1267–1274

11. Ogg GS XJ, Sebastian B (1998) Quantitation of HIV-1-specific cytotoxic T lymphocytes and plasma load of viral RNA. Science 279:2103–2106

12. Klein MR, van Baalen CA, Holwerda AM, Kerkhof Garde SR, Bende RJ, Keet IP, Eeftinck-Schattenkerk JK, Osterhaus AD, Schuitemaker H, Miedema F (1995) Kinetics of Gag-specific cytotoxic T lymphocyte responses during the clinical course of HIV-1 infection: a longitudinal analysis of rapid progressors and long-term asymptomatics. J Exp Med 181:1365–1372

13. Harrer T, Harrer E, Kalams SA, Barbosa P, Trocha A, Johnson RP, Elbeik T, Feinberg MB, Buchbinder SP, Walker BD (1996) Cytotoxic T lymphocytes in asymptomatic long-term nonprogressing HIV-1 infection. Breadth and specificity of the response and relation to in vivo viral quasi-species in a person with prolonged infection and low viral load. J Immunol 156:2616–2623

14. Carmichael A, Jin X, Sissons P, Borysiewicz L (1993) Quantitative analysis of the human immunodeficiency virus type 1 (HIV-1)-specific cytotoxic T lymphocyte (CTL) response at different stages of HIV-1 infection: differential CTL responses to HIV-1 and Epstein-Barr virus in late disease. J Exp Med 177:249–256

15. Brodie SJ LD, Patterson BK (1999) In vivo migration and function of transferred HIV-1-specific cytotoxic T cells. Nat Med 5:34–41

16. Nash RA, Sullivan Pepe M, Storb R, Longton G, Pettinger M, Anasetti C, Appelbaum FR, Bowden RA, Deeg HJ, Doney K, Martin PJ, Sullivan KM, Sanders J, Witherspoon RP (1992) Acute graft-versus-host disease: analysis of risk factors after allogeneic bone marrow transplantation and prophylaxis with cyclosporine and prednisone. Blood 80:1838–1846

17. Horowitz MM, Gale RP, Sondel PM, Goldman JM, Kersey J, Kolb HJ, Rimm AA, Ringden O, Rozman C, Speck B (1990) Graft-versus-leukemia reactions after bone marrow transplantation. Blood 75:555–562

18. Collins RH, Shpilberg O, Drobyski W, Porter DL, Giralt S, Champlin R, Goodman SA, Wolff SN, Hu W, Verfaille C, List A, Dalton W, Ognoskie N, Chetrit A, Antin JH, Nemunaitis J (1997) Donor leukocyte infusions in 140 patients with relapsed malignancy after allogeneic bone marrow transplantation. J Clin Oncol 15:433–444

19. Warren EH, Greenberg PD, Riddell SR (1998) Cytotoxic T-lymphocyte-defined human minor histocompatibility antigens with a restricted tissue distribution. Blood 91:2197–2207

20. Dolstra H, Fredrix H, Preijers F, Goulmy E, Figdor CG, de Witte TM, van de Wiel-van Kemenade E (1997) Recognition of a B cell leukemia-associated minor histocompatibility antigen by CTL[1]. J Immunol 158:560–565

21. de Bueger M, Bakker A, van Rood JJ, Van der Woude F, Goulmy E (1992) Tissue distribution of human minor histocompatibility antigens. Ubiquitous versus restricted tissue distribution indicates heterogeneity among human cytotoxic T lymphocyte-defined non-MHC antigens. J Immunol 149:1788–1794

22. Goulmy E (1997) Human minor histocompatibility antigens: new concepts for marrow transplantation and adoptive immunotherapy. Immunol Rev 157:125–140

23. Wang W, Meadows LR, den Haan JM, Sherman NE, Chen Y, Blokland E, Shabanowitz J, Agulnik A, Hendrickson RC, Bishop CE, Hunt DF, Goulmy E, Engelhard VH (1995) Human H-Y: a male-specific histocompatibility antigen derived from the SMCY protein. Science 269:1588–1590

24. Meadows L, Wang W, den Haan JM, Blokland E, Reinhardus C, Drijfhout JW, Shabanowitz J, Pierce R, Agulnik AI, Bishop CE, Hunt DF, Goulmy E, Engelhard VH (1997) The HLA-A→1-restricted H-Y antigen contains a posttranslationally modified cysteine that significantly affects T cell recognition. Immunity 6:273–281

25. Goulmy E, Schipper R, Pool J, Blokland E, Frederik Falkenburg JH, Vossen J, Gratwohl A, Vogelsang GB, van Houwelingen HC, van Rood JJ (1996) Mismatches of minor histocompatibility antigens between HLA-identical donors and recipients and the development of graft-versus-host disease after bone marrow transplantation. N Engl J Med 334:281–285

26. van der Harst D, Goulmy E, Falkenburg JH, Kooij-Winkelaar YM, van Luxemburg-Heijs SA, Goselink HM, Brand A (1994) Recognition of minor histocompatibility antigens on lymphocytic and myeloid leukemic cells by cytotoxic T-cell clones. Blood 83:1060–1066

27. Falkenburg JH, Goselink HM, van der Harst D, van Luxemburg-Heijs SA, Kooy-Winkebar YM, Faber LM, de Kroon J, Brand A, Fibbe WE, Willemze R (1991) Growth inhibition of clonogenic leukemic precursor cells by minor histocompatibility antigen-specific cytotoxic T lymphocytes. J Exp Med 174:27–33

28. Bonnet D, Dick JE (1997) Human acute myeloid leukemia is organized as a hierarchy that originates from a primitive hematopoietic cell. Nat Med 3:730–737

29. Morse MC, Bleau G, Dabhi VM, Hetu F, Drobetsky EA, Lindahl KF, Perreault C (1996) The COI mitochondrial gene encodes a minor histocompatibility antigen presented by H2-M3. J Immunol 156:3301–3307

30. Bhuyan PK, Young LL, Lindahl KF, Butcher GW (1997) Identification of the rat maternally transmitted minor histocompatibility antigen. J Immunol 158:3753–3760

31. Gubarev MI, Jenkin JC, Leppert MF, Buchanan GS, Otterud BE, Guilbert DA, Beatty PG (1996) Localization to chromosome 22 of a gene encoding a human minor histocompatibility antigen. J Immunol 157:5448–5454

32. Hunt DF, Henderson RA, Shabanowitz J, Sakaguchi K, Michel H, Sevilir N, Cox AL, Appella E, Engelhard VH (1992) Characterization of peptides bound to the class I MHC molecule HLA-A2.1 by mass spectrometry. Science 255:1261–1263

33. den Haan JMM, Sherman NE, Blokland E, Huczko E, Koning F, Wouter Drijfhout J, Skipper J, Shabanowitz J, Hunt DF, Engelhard VH, Goulmy E (1995) Identification of a graft versus host disease-associated human minor histocompatibility antigen. Science 268:1476–1480

34. den Haan JMM, Meadows LM, Wang W, Pool J, Blokland E, Bishop TL, Reinhardus C, Shabanowitz J, Offringa R, Hunt DF, Engelhard VH, Goulmy E (1998) The minor histocompatibility antigen HA-1: a diallelic gene with a single amino acid polymorphism. Science 279:1054–1057

35. Boon T, Cerottini JC, Van den Eynde B, van der Bruggen P, Van Pel A (1994) Tumor antigens recognized by T lymphocytes. Annu Rev Immunol 12:337–365

36. Dolstra HFH, Maas F (1999) A human minor histocompatibility antigen specific for B cell acute lymphoblastic leukemia. J Exp Med 189:301–308

37. Clift RA, Buckner CD, Appelbaum FR, Bearman SI, Petersen FB, Fisher LD, Anasetti C, Beatty P, Bensinger WI, Doney K, Hill RS, McDonald GB, Martin P, Sanders J, Singer J, Stewart P, Sullivan KM, Witherspoon R, Storb R, Hansen JA, Thomas ED (1990)

Allogeneic marrow transplantation in patients with acute myeloid leukemia in first remission: a randomized trial of two irradiation regimens. Blood 76:1867–1871

38. Bonini C, Ferrari G, Verzeletti S, Servida P, Zappone E, Ruggieri L, Ponzoni M, Rossini S, Mavilio F, Traversari C, Bordignon C (1997) HSV-TK gene transfer into donor lymphocytes for control of allogeneic graft-versus-leukemia. Science 276:1719–1724

39. Riddell SR, Elliot M, Lewinsohn DA, Gilbert MJ, Wilson L, Manley SA, Lupton SD, Overell RW, Reynolds TC, Corey L, Greenberg PD (1996) T cell mediated rejection of gene-modified CD8+ HIV-specific cytotoxic T lymphocytes in HIV infected subjects. Nat Med 2:216–223

Adoptive Transfer of Polyclonal, EBV-Specific Cytotoxic T-Cell Lines for the Prevention and Treatment of EBV-Associated Malignancies

Cliona M. Rooney[1], M. Helen Huls[1], Richard A. Rochester[2], Malcolm K. Brenner[1], and Helen E. Heslop[1]

Summary. There is increasing interest in using adoptive transfer of antigen-specific cytotoxic T lymphocytes (CTL) for the treatment of cancer. The challenges facing this strategy include identification of tumor antigens against which to target CTLs, the development of methods to activate and expand tumor-specific CTLs in vitro, and ensuring the in vivo survival and function of the CTLs in the face of multiple antiimmune response strategies employed by viruses and tumors. Epstein–Barr virus (EBV) provides ideal model systems in which to test the safety and efficacy of infusions of virus-specific CTLs for the treatment of tumors with and without immune evasion strategies. EBV-transformed B-cell lines (LCLs) are excellent antigen-presenting cells (APCs) that express nine virus-encoded proteins and can be used to stimulate and expand EBV-specific CTL lines from most donors in vitro. EBV-associated lymphomas most commonly occur in patients who are severely immunosuppressed and hence do not use immune evasion strategies. We have shown that these tumors can be treated effectively with infusions of virus-specific CTLs. The tumors occurring in patients who are immunocompetent or only mildly immunosuppressed do employ immune evasion strategies and these provide a model for the use of CTLs that have been genetically modified to persist and function under adverse conditions.

Key words. Adoptive immunotherapy, Epstein–Barr virus (EBV), Cytotoxic T lymphocyte (CTL), Clinical immunology, Tumor-specific CTL

[1]Center for Cell and Gene Therapy, Baylor College of Medicine, 6621 Fannin Street, Houston, TX 77030, USA
[2]Department of Hematology and Oncology, St. Jude Children's Research Hospital, 332 North Lauderdale, Memphis, TN 38105, USA

45

Introduction

Adoptive Immunotherapy

Immunotherapeutic strategies may be useful when a tumor or virus-infected cell fails to activate an effective immune response in the host. This may occur for a number of reasons. First, the tumor cell may fail to express an antigen that is perceived as foreign. Second, a tumor cell that expresses a foreign antigen may be a poor antigen-presenting cell (APC) because it fails to express the costimulatory molecules required for activating an immune response (Fig. 1). Third, the tumor cell may have mechanisms to inhibit the appropriate immune response. Fourth, the host may be immunosuppressed. The goal of immunotherapy is to overcome the deficits of the host or the tumor itself and to activate an effective immune response to the tumor. The cytotoxic T lymphocyte (CTL) arm of the cellular immune response is thought to be the most important defense against tumors and virus-infected cells. Although the foreign antigens recognized by CTL are commonly viral proteins expressed in infected cells, CTL epitopes may also be provided by aberrantly expressed self-proteins. For example, self-proteins that are overexpressed, aberrantly spliced, or contain deletions or mutations may be perceived as foreign. There is also some interest in breaking tolerance to normal self-antigens for

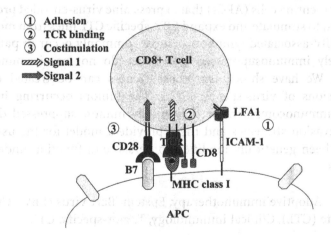

FIG. 1. A good antigen-presenting cell (APC) (dendritic cell, macrophage, or B cell) expresses an immunogenic peptide in the MHC class I molecule. This complex is recognized by a CD8+ CTL precursor by its T cell receptor and by its CD8 molecule. This union is strengthened by the presence of adhesion molecules such as ICAM1 and its ligand LFA1. If a costimulatory signal is also provided by B7 on the target cell which interacts with CD28 on the CTLp, then the CTL is activated

immunotherapeutic purposes. In this chapter, we concentrate on Epstein–Barr virus (EBV) -specific CTLs as immunotherapy for EBV-associated malignancies in which foreign CTL epitopes are provided by virally encoded proteins.

The Cytotoxic T Cell

HLA class I-restricted CD8+ CTL are the best studied mediators of tumor regression. They recognize short peptides, usually derived from endogenously synthesized foreign antigens, that are degraded in the cytosol, and transported to the endoplasmic reticulum for loading onto nascent HLA class I molecules [1]. Exogenous proteins may also enter the HLA class I pathway if they can access the cytosol. Viruses have mechanisms to enter the cytosol where virion proteins can then be targeted to the class I pathway. Thus virion proteins from cytomegalovirus and adenovirus activate virion protein-specific CD8+ CTL responses, without a requirement for virus gene expression [2,3]. Professional APCs may also target phagocytosed particles to the class I pathway in a phenomenon known as cross-priming [4]. CD4+ CTLs that recognize peptides in association with class II also exist, although their role in antitumor immunity remains obscure. CD4+ CTLs recognize exogenous or endogenous proteins that are processed endocytically and loaded onto newly synthesized or recycling HLA class II molecules [5].

For immunotherapeutic purposes, CTLs may be activated in vivo or ex vivo. In vivo strategies involve immunization with DNA, tumor vaccines, or antigen-loaded dendritic cells. Ex vivo strategies involve exposing T cells to tumor or viral antigens expressed on APC and expanding them in T-cell growth factors in vitro. Although the ex vivo approach is more costly in terms of time, effort, and expertise required to grow CTLs for patient infusion, it may be the only option in cases in which the patient is immunosuppressed or the tumor secretes inhibitory factors. In these cases it is often possible to activate and expand antigen-specific CTL precursors in a culture environment that is conducive to CTL growth. The expanded CTLs may then be returned to the patient when sufficient numbers have been obtained. There are additional advantages to the ex vivo approach. The phenotype and function of the CTLs can be determined before infusion, antitumor activity can be ensured and antihost activity can be excluded. CTL numbers can be controlled and repeat infusions given if required. Finally, transfer of marker genes into CTL lines allows the function and persistence of CTLs in vivo to be followed [6], transfer of suicide genes allows the in vivo destruction of the CTLs should they prove toxic [7], and transfer of functional genes may improve the activity of CTLs once infused [8]. We have used the ex vivo approach for the prevention and treatment of EBV-related malignancies in stem cell recipients, because they are severely immunosuppressed, and in patients with relapsed EBV-positive Hodgkin disease, because the tumor secretes inhibitory factors.

The Epstein–Barr Virus: Latency, Tumorigenesis, and Subversion of the Host Immune Response

The Epstein–Barr virus provides an excellent model system in which to study the safety and biological efficacy of adoptively transferred CTLs. A majority of individuals are infected with EBV in early childhood and the virus then persists for life in B lymphocytes and epithelial cells. EBV efficiently transforms B cells in vitro into immortalized (lympho blastoid) cell lines (LCLs) and is associated with malignancies, not only of B cells and epithelial cells but also of T cells, natural killer (NK) cells, and muscle [9,10]. All EBV-positive malignancies are associated with the latent life cycle of the virus, during which it expresses only a small proportion of its genome. The differential regulation of these genes is one of many strategies the virus employs to evade the immune response. In the most immunogenic tumors, ten proteins and two small RNAs are expressed: the EBV nuclear antigens (EBNAs) 1, 2, 3A, 3B, 3C, and LP; the latent membrane proteins (LMPs) 1, 2a, and 2b; the cytosolic protein (RK-BARFO) derived from the *Bam*HI A rightward transcripts; and two small nonpolyadenylated RNAs (EBERs). EBNA-1 is required for maintenance of the virus episome in dividing cells and has tumorigenic activity in transgenic mice [12]. EBNA-2 is a transforming protein that binds the cellular transcription factor, RBPJ-κ, and controls the expression of both viral and cellular genes such as c-*fgr*, CD21, and CD23 [13,14]. EBNAs 3A, 3B, and 3C and EBNA-LP regulate EBNA-2 [15,16]. LMP-1 is a transforming protein that functions as a constitutively active, ligand-independent signaling molecule. It is a member of the tumor necrosis factor receptor (TNFR) superfamily, and interacts with TNFR-associated proteins (TRAFs) that activate NFκB and epidermal growth factor receptor (EGFR) and modulate cell growth and apoptosis [17,18]. LMP-2 blocks signaling through the B-cell receptor and prevents apoptosis, providing a survival signal [19]. Roles for the *Bam*HI A transcripts and the EBER small RNAs have not yet been determined, but as they are expressed in all virus-associated tumors they are likely to have important functions.

Three types of latency have so far been described in EBV-positive tumors (Table 1). Burkitt's lymphoma (BL) and gastric carcinoma express type 1 latency, which involves the expression of only four EBV genes, EBNA-1, RK-BARFO, and the two EBERs. Although both EBNA-1 and RK-BARFO contain predicted CTL epitopes, CTLs raised against these peptide epitopes do not kill EBV-infected target cells [20,21]. Thus, both downregulation of viral antigens and resistance to CTLs specific for those antigens expressed provide immune

TABLE 1. Patterns of Epstem–Barr virus (EBV) latency

Latency	EBV genes expressed
0	LMP2a
1	EBNA1, BARFO, EBERs
2	EBNA1, BARFO, EBERs, LMP1, LMP2
3	EBNA1, BARFO, EBERs, LMP1, LMP2, EBNA2, EBNA3a, EBNA3b, EBNA3c, EBNA-LP

evasion strategies for BL cells. BL cells have additional mechanisms to evade immune responses. They downregulate their expression of HLA class I alleles, of immune costimulator molecules, and of the transporter (TAP) molecules required for loading of peptides onto HLA class I molecules [22–24]. The presence of multiple antiimmune response mechanisms suggests that on its own none is 100% effective.

Type 2 latency, seen in Hodgkin's lymphoma, nasopharyngeal carcinoma, NK and T-cell lymphomas, and other mucosal carcinomas, involves the expression of two additional membrane proteins, LMP-1 and LMP-2. Both proteins encode effective, if minor, CTL epitopes, so that at first glance, survival of cells

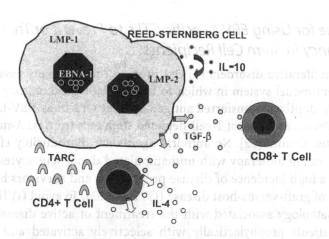

FIG. 2. The malignant Reed-Sternberg cell of Hodgkins lymphoma expresses the immunogenic viral antigens *LMP1* and *LMP2* that have been shown to be immunogenic. They also express HLA class I and class II molecules, as well as adhesion and costimulatory molecules. They should therefore be good APCs. However, they secrete *IL-10*, which prevents cross-priming: *TGFβ*, which directly inhibits CTL function; and *TARC*, a chemokine attractant for CD4+ T helper cells of the TH2 phenotype. Thus they inhibit the recruitment, activation, and function of virus-specific CTLs

expressing type 2 latency is less easy to explain. This is particularly puzzling in the case of Hodgkin lymphoma because Reed–Sternberg cells express high levels of immune costimulatory molecules and have intact antigen-processing machinery and should therefore be good APC and good targets [25,26]. However, Reed–Sternberg cells secrete inhibitory cytokines such as transforming growth factor-β (TGF-β) and IL-10 and the TH2 chemokine, TARC [27–30]. TGF-β directly inhibits CTLs, and IL-10 inhibits the activation of professional APC and may thus prevent cross-priming and inflammation. TARC specifically recruits TH2-type CD4 cells that are inhibitory to the TH1 cells required for CTL activation. The Hodgkin tumor cell may thus protect itself from CTLs by creating a TH2 environment (Fig. 2). Other tumors expressing type 2 latency may employ similar mechanisms.

Type 3 latency involves the expression of all the EBV latency-associated proteins, some of which are highly immunogenic. CTL epitopes in the EBNA-3 proteins, expressed only in type 3 latency, dominate the cellular immune response to EBV in almost all individuals. Cells expressing type 3 latency do not evade the immune response and are seen only in severely immunocompromised individuals [31]. These cells cause the lymphoproliferative disorders of individuals immunocompromised by organ or stem cell transplantation, congenital immunodeficiency, or HIV infection. They are phenotypically identical to the lymphoblastoid cell lines (LCLs) created by infection of B lymphocytes with EBV in vitro.

Rationale for Using EBV-Specific CTLs to Prevent or Treat EBV Malignancy in Stem Cell Recipients

Lymphoproliferative disorders (LPDs) in stem cell recipients have provided an excellent model system in which to test the biological efficacy of ex vivo expanded, adoptively transferred antigen-specific CTL lines. EBV-LPD occurs in up to 20% of recipients of T-cell-depleted stem cells from HLA-mismatched or unrelated donors [32]. No antiviral agents are reproducibly effective for EBV-LPD. Immunotherapy with unmanipulated donor leukocytes is associated with a high incidence of disease progression and survivors have a high incidence of graft-versus-host disease (GVHD) [33]. To avoid GVHD and the immunopathology associated with the treatment of active disease, we have treated patients prophylactically with selectively activated and expanded EBV-specific CTLs. EBV-specific CTL lines are readily established from the majority of stem cell donors, most of whom are persistently infected with EBV and therefore carry a high frequency of circulating EBV-specific CTL precursors. LCLs are easy to establish from normal donors, reproducibly activate EBV-specific CTL lines from seropositive donors and provide a continuous source of excellent APC.

Materials and Methods

Production and Concentration of EBV

B95-8 cells were derived from our master cell bank. These cells have been screened for adventitious viruses, including the simian retrovirus that contaminates some cultures of B95-8, and are negative for mycobacteria, bacteria, and fungi. They have the karyotype of the original B95-8 marmoset cell line (one large metacentric chromosome, thought to be derived from an acrocentric chromosome 16). We suspended 800×10^6 cells in 800 ml of RPMI 1640 (GIBCO-BRL) containing L-glutamine and 5% fetal calf serum (Hyclone) in 175-cm^2 flasks or similar vessels. The cells are then incubated for 7 days at 37°C in 5% CO_2 in air. The supernatant is harvested and cells removed by centrifugation at $450 \times g$ for 20 min in 200-ml bottles. The supernatant is filtered through a 0.8-μM bottle top filter, then concentrated 50 fold by circulation through a hollow fiber filter (Pellicon XL polyetherone sulfate filter with a 300K (Millipore PXB300C50), using a Labscale TM TFF system (pump, tubing, and reservoir) from Millipore, exactly according to the manufacturer's instructions. The virus is then filtered through a 0.45-μM Millipore filter, and stored in aliquots of 200 μl at −80°C. Using this method, no virus was detected in the eluate, and the concentrated virus was at least 50 fold more active than unconcentrated virus. Note that the virus is sensitive to temperatures above −70°C and cannot be freeze-thawed without loss of titer.

To test the virus titer, 6×10^5 peripheral blood mononuclear cells (PBMC) in 100 μl complete medium (RPMI 1640 [GIBCO] containing L-glutamine and 10% fetal calf serum [Hyclone]) are incubated with 10-fold dilutions of virus (100 μl) from 10^{-1} to 10^{-6} in the presence of 1 μg/ml cyclosporin A. Virus and cells are incubated in triplicate in 96-well flat-bottomed plates. A good concentrated virus should transform at a dilution of 10^{-5} to ensure rapid preparation of LCL.

Generation of EBV-Transformed B-Cell Lines for Use as APC

For this step, 5×10^6 PBMCs are resuspended in 200 μl of concentrated supernatant from the B95-8 virus producer line. The volume is made up to 2 ml with complete medium containing 2 μg/ml cyclosporin A; 200 μl is added to 5 wells and 100 μl to 10 wells of a 96-well flat-bottomed plate. All wells are made up to 200 μl with medium/CSA. Sterile water is aliquoted into empty wells to prevent evaporation, and the plate is incubated for 2–4 weeks, feeding weekly by a half change of medium. After 2–3 weeks, if large clumps are expanding, sample wells can be combined into a 2-ml well. If this well expands, remaining wells can be transferred. This transfer is critical because LCLs will be lost if they are expanded too quickly. When the LCL are growing well in 2-ml wells, they can be transferred to 25-cm² flasks. Several vials of LCLs should be cryopreserved before they are used to initiate CTL lines. After about 2 weeks of culture, the LCLs are maintained in 100 µM acyclovir to prevent the production of infectious virus. Samples are tested for sterility and mycoplasma before CTL initiation.

Generation of EBV-Specific CTL Lines

Fresh or frozen donor PBMCs are resuspended at 2×10^6 cells/ml of complete medium; 1 ml of PBMC is aliquoted per 2-ml well. The autologous LCL is irradiated with 40 Gy, washed once to remove superoxide radicals, counted, and resuspended at 5×10^4 cells/ml; 1 ml LCL is then added per well of PBMCs. An additional well contains 2 ml LCL alone to control for irradiation. The culture is fed with a half change of medium on day 7, then harvested and counted on day 10. Dead cells are removed on lymphocyte separation medium (Nycomed) if necessary, and 10^6 responder cells and 2.5×10^5 irradiated autologous LCL are plated in 24 well plates in cells in 2 ml of complete medium. IL-2 at 40–80 units/ml is added on day 14, and thereafter the cultures are restimulated weekly with irradiated LCL (4:1 ratio) and twice weekly with IL-2. When sufficient cell numbers for patient treatment are obtained, the cells are safety tested (see Table 2) and frozen using a controlled rate freezer.

Testing the Cytotoxic Activity of CTL Lines

CTL lines are tested in a standard chromium release assay. Target cells include the autologous LCL, an HLA class I-mismatched LCL, a lymphokine-activated killer (LAK) -sensitive T-cell lymphoma, HSB-2, and patient-derived (pretransplant) phytohemagglutinin (PHA) blasts. Antibodies to HLA class I and II are added to the autologous LCL to determine the class restriction. CTLs and chromium-labeled targets are incubated in triplicate for 4 h at effector/target ratios of 40:1, 20:1, 10:1, and 5:1. Only killing of patient PHA blasts of greater than 10% excluded the CTL line from infusion.

TABLE 2. Safety testing of cytotoxic T lymphocyte (CTL) lines

Test	Required result
Sterility	Negative for growth of fungus and bacteria
Endotoxin	<5 IU
Mycoplasma	Negative
HLA type	Donor HLA class I allotype
Phenotype	<1% CD19+ B cells
Cytotoxicity	<10% killing of patient-derived normal cells (phytohemagglutinin-stimulated PBMCs)
	Killing of donor LCL

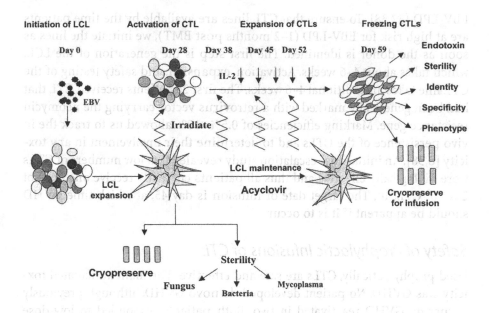

Fig. 3. Generation of EBV-specific cytotoxic T cell lines. EBV-transformed B cell lines (LCLs) are derived by infection of PBMC with a laboratory strain of EBV, B95-8. LCLs are irradiated and used to stimulate autologous PBMC. The responding T cells are restimulated weekly with irradiated LCLs, and expanded by twice weekly activation with 80 units per ml of IL-2. When T cell numbers are sufficient for patient treatment, they are safety-tested as indicated and cryo-preserved. When safety test results are completed and appropriate, the CTLs are released for clinical use

The CTL Phenotype

The CTL line is incubated with fluorochrome-labeled antibodies to CD3, CD4, CD8, CD16, CD56, CD19, TCR αβ, and TCR γδ. CD4+ T cells ranged from 2% to 98%, as did CD8+ T cells; CD56+CD3+ cells ranged from 1% to 49%. Because CD56+ T cells are not associated with GVHD, only the presence of more than 1% CD19+ cells excluded the CTL line from infusion [35,36]. The entire process of CTL generation is portrayed in Fig. 3.

Results

Establishment of CTL Lines

EBV-specific CTL lines were established from all but 3 of more than 100 donors. Two of these were seronegative and 1 had recently seroconverted. Since 1993 we have infused donor-derived, EBV-specific CTL lines into more than 60 bone marrow transplant (BMT) recipients as prophylaxis for

EBV-LPD [37,38]. To ensure that CTL lines are available by the time patients are at high risk for EBV-LPD (1–2 months post BMT), we initiate the lines as soon as the donor is identified. The first step is the generation of the LCL, which takes about 4–6 weeks. Activation, expansion, and safety testing of the CTL line takes an additional 4–6 weeks. The first 26 patients received CTL that had been genetically marked with a retrovirus vector carrying the neomycin resistance gene. Marking efficiencies of 0.5%–10% allowed us to track the in vivo persistence of the CTLs and to determine their involvement in any toxicity [6,38]. An initial dose escalation study revealed that low numbers of CTLs were biologically effective and thus all patients currently receive one dose of 2×10^7 CTL/m^2. The target date of infusion is day 45, at which time GVHD should be apparent if it is to occur.

Safety of Prophylactic Infusions of CTL

Used prophylactically, CTLs are safe and effective. The main anticipated toxicity was GVHD. No patient developed de novo GVHD, although previously occurring GVHD reactivated in two. Both patients responded to low-dose steroids. No other toxicities could be associated with CTL infusions.

Persistence of CTLs

Amplification of marker DNA from EBV-specific CTLs reactivated from patients at serial times after infusion showed that the infused CTL lines expanded up to four logs in vivo and then persisted for as long as 5 years after infusion [6,38,39]. Because endogenous EBV-specific CTLs return at about 8–9 months post BMT, this persistence is easily long enough for protection. Marking studies also showed that the CTLs could expand in vivo in response to EBV reactivation; a rise in EBV DNA levels in one patient 18 months post infusion was accompanied by reappearance of the marker gene [6].

Reconstitution of Immune Responses to EBV

Before CTL infusion, we were unable to detect EBV-specific CTL precursors. Within 1 week of infusion, low-level activity could be detected, and within 4 weeks the levels of immunity were within the range seen in normal individuals [6,39].

CTLs Reduce a High Virus Load

About 20% of the patients developed a high virus load, as measured by amplification of EBV DNA in peripheral blood. High EBV DNA levels are frequently accompanied by fevers and lymphadenopathy and are associated with an extremely high risk for EBV-LPD [34]. Infusion of CTL into patients with a high virus load uniformly resulted in a dramatic drop in virus load to low or undetectable levels [38,39].

EBV-Specific CTLs Have Antitumor Activity

The most important result of the study was that none of the patients who received prophylactic CTL developed EBV-LPD, in contrast with about 12% of controls [38].

Further evidence for antitumor effects came from three patients who had not received CTLs and developed frank lymphoma. These three received CTLs as treatment, and two of the three achieved complete remission [38]. The second case clearly demonstrated that EBV-specific CTLs home to tumor sites, then accumulate or expand to effect lymphoma regression. Comparison of tumor biopsy material taken before and 10 days after CTL infusion showed that CD20+ tumor B cells were replaced almost entirely by CD3+ T cells [38]. In situ PCR demonstrated neo-marked CTLs in the tumor tissue at levels more than 100 fold those seen in peripheral blood.

Pitfalls Associated with the Use of CTLs as Treatment

The three patients who received CTLs as treatment also illustrated two of the pitfalls that can result from treating bulky disease. First, if the tumor occurs in a sensitive anatomical location, the inflammatory response can be damaging and, second, mutation of important CTL epitopes becomes increasingly likely with tumor size. While the first patient achieved complete remission without a problem, the second patient, who presented with airway obstruction from bulky disease in the nasopharynx, developed increased swelling after CTL infusion, requiring intubation and tracheotomy [38]. He also developed ulceration of the soft palate and gut, illustrating the damage that a CTL-mediated inflammatory response can generate. The patient eventually recovered and remains well over 3 years after CTL infusion.

The third patient also presented with bulky disease involving the nasopharynx and lung [40]. She received CTLs but the tumor progressed and she died of respiratory failure 24 days after CTL infusion. EBV-transformed B-cell lines grew rapidly from patient peripheral blood before and 7 days after CTL infusion. Comparative analysis of tumor virus DNA with DNA from the patient-derived B-cell lines and B95-8 DNA revealed a deletion in the EBNA3B gene in the patient's EBV. This deletion removed two A11-restricted epitopes that are immunodominant in all A11-positive Caucasian individuals. Analysis of this A11-positive donor CTL line showed that it also was dominated by these two EBNA 3B epitopes. No other epitope specificity could be identified from peptides predicted from the literature to sensitize targets to killing through the donor HLA class I allotype [31]. The presence of the deletion in tumor cells grown before CTL infusion showed that the deletion had occurred before the CTL infusion. It is not yet clear whether the deleted virus arose as a minor strain in the donor or was originally present in the recipient.

Conclusions

The use of EBV-specific CTL lines is a safe and effective strategy to prevent EBV-associated disease after stem cell transplant. The CTL lines rapidly reconstitute immune responses to EBV and persist for years after infusion. Prophylactic CTL infusions are safe and provide immunological protection against EBV-LPD. However, CTLs as therapy should be used with caution. Inflammatory responses can be damaging because of the aggressive nature of the tumor and its preference for respiratory tissues. Escape mutants may also be a problem because even polyclonal EBV-specific CTL lines may be specific for only a few viral epitopes, and with a large tumor burden there is ample opportunity for mutation.

Avoiding the Pitfalls of Immunotherapy

To avoid damaging inflammatory responses to bulky tumors, it is clearly advantageous to treat early or minimal residual disease or to treat prophylactically. The strategy employed depends on the nature of the tumor. Prophylaxis is our preferred strategy and has been effective. Nevertheless, regular virus load monitoring is important in the management of patients who may be ineligible for CTL for any reason. To prevent tumors from mutating to avoid the immune response, one strategy is to target epitopes that are essential for the transformed phenotype. However, it seems unlikely that such essential domains will regularly provide effective CTL epitopes. A second strategy is to use polyclonal CTLs or clones that target multiple epitopes. Although the EBV-specific CTL lines we generate are polyclonal and rarely show a bias in their use of T cell receptor Vβ gene usage, they are often monoclonal or oligoclonal in specificity. Such was the case with our donor CTL line that failed.

Other Uses for EBV-Specific CTL Lines

EBV-specific CTL lines may be used to treat EBV-LPD arising in other circumstances. Three groups of organ transplant recipients are at particularly high risk: seronegative recipients of EBV-carrying organs, recipients of organs such as gut that carry a high B-cell load, and recipients who experience repeat rejection episodes and receive prolonged, intensive immunosuppression. Surprisingly it has been possible to generate EBV-specific CTL lines even from recipients with lymphoproliferative disease [9] (and our observations), possibly because EBV-specific CTL precursors exist in the circulation but are unable to expand in vivo in the presence of immunosuppressive drugs. If these precursors are removed into a supportive culture environment, they can respond to activation and proliferation signals. In the case of seronegative recipients, it may be possible to activate CTL using dendritic cells as APCs because these potent cells are able to activate CTLs from naive precursors in vitro. Used prophylactically in a seronegative recipient, CTLs may prevent primary infection from developing into lymphoma.

In chronically immunosuppressed individuals, infused CTL may be inactivated over time and repeat CTL infusions may be required. Virus load should be a good indicator of CTL function and allow timely intervention. CTL prepared in advance in patients identified as having high risk status can be infused when the virus load increases. Some patients have persistently high levels of EBV DNA, but do not develop lymphoma. This increased virus-driven lymphoproliferation will increase the likelihood of oncogenic mutations, which may eventually lead to the development of lymphomas that are less dependent on EBV and therefore will be less susceptible to immunotherapeutic approaches.

Patients with immunodeficiency disorders are also at increased risk of EBV-LPD. Again, surprisingly, it has been possible to generate EBV-specific CTLs from some of these individuals, and the CTL have produced good results in vivo. EBV-positive Hodgkin's disease is another candidate for EBV-specific CTL. Five patients who received EBV-specific CTL had temporary improvements, including increases in EBV-specific CTL precursor frequency, reduction in high virus load, resolution of type B symptoms, and stabilization of disease [42]. However, all patients eventually died. Current improvements include the generation of CTL lines that are specific for the limited range of viral antigens that are expressed in Reed–Sternberg cells. Dendritic cells, transduced with viral vectors that express individual EBV proteins, are effective antigen-presenting cells, even in naive or nonresponder individuals [43,44]. Future developments may involve transduction of antigen-specific CTLs with genes that facilitate CTL function in a TH2 environment.

Acknowledgments. This work was supported in part by NIH grants CA61384, CA74126 and Cancer Center Support CORE Grant 21765, the American Lebanese Syrian Associated Charities (ALSAC), and the Department of Pediatrics, Baylor College of Medicine. We thank Belinda Rossitter for editing the manuscript.

58   C.M. Rooney et al.

References

1. Jondal M, Schirmbeck R, Reimann J (1996) MHC class-I restricted CTL responses to exogenous antigen. Immunity 5:295–302
2. Gilbert MJ, Riddell SR, Plachter B, Greenberg PD (1996) Cytomegalovirus selectively blocks antigen processing and presentation of its intermediate-early gene product. Nature (Lond) 383:720–722
3. Smith CA, Woodruff LS, Kitchingman GR, Rooney CM (1996) Adenovirus-pulsed dendritic cells stimulate human virus-specific T-cell responses in vitro. J Virol 70:6733–6740
4. Inaba K, Turley S, Yamaide F, Iyoda T, Mahnke K, Inaba M, Pack M, Subklewe M, Sauter B, Sheff D, Albert M, Bhardwaj N, Mellman I, Steinman RM (1998) Efficient presentation of phagocytosed cellular fragments on the major histocompatibility complex class II products of dendritic cells. J Exp Med 188:2163–2173
5. Lanzavecchia A (1996) Mechanisms of antigen uptake for presentation. Curr Opin Immunol 8:348–354
6. Heslop HE, Ng CYC, Li C, Smith CA, Loftin SK, Krance RA, Brenner MK, Rooney CM (1996) Long-term restoration of immunity against Epstein-Barr virus infection by adoptive transfer of gene-modified virus-specific T lymphocytes. Nat Med 2:551–555
7. Bonini C, Ferrari G, Verzeletti S, Servida P, Zappone E, Ruggieri L, Ponzoni M, Rossini S, Malvilio F, Traversari C, Bordignon C (1997) HSV-TK gene transfer into donor lymphocytes for control of allogeneic graft versus leukemia. Science 276:1719–1724
8. Hwu P, Shafer GE, Treisman J, Schindler DG, Gross G, Cowherd R, Rosenberg SA, Eshhar Z (1993) Lysis of ovarian cancer cells by human lymphocytes redirected with a chimeric gene composed of an antibody variable region and the Fc receptor gamma chain. J Exp Med 178:361–366
9. McClain KL, Leach CT, Jenson HB, Joshi VV, Pollock BH, Parmley RT, DiCarlo TJ, Chadwick EG, Murphy SB (1995) Association of Epstein-Barr virus with leiomyosarcomas in children with AIDS [see comments]. N Engl J Med 332:12–18
10. Su IJ, Lin KH, Chen CJ, Tien HF, Hsieh HC, Lin DT, Chen JY (1990) Epstein-Barr virus-associated peripheral T-cell lymphoma of activated CD8 phenotype. Cancer (Phila) 66:2557–2562
11. Yates JL, Warren N, Sugden B (1985) Stable replication of plasmids derived from Epstein-Barr virus in various mammalian cells. Nature 313:812–815
12. Wilson JB, Levine AJ (1992) The oncogenic potential of Epstein-Barr virus nucear antigen 1 in transgenic mice. Curr Top Microbiol Immunol 182:375–384
13. Calender A, Cordier M, Billaud M, Lenoir GM (1990) Modulation of cellular gene expression in B lymphoma cells following in vitro infection by Epstein-Barr virus (EBV). Int J Cancer 46:658–663
14. Henkel T, Ling PD, Hayward SD, Peterson MG (1994) Mediation of Epstein-Barr virus EBNA2 transactivation by recombination signal-binding protein J kappa. Science 265:92–95
15. Harada S, Kieff E (1997) Epstein-Barr virus nuclear protein LP stimulates EBNA-2 acidic domain-mediated transcriptional activation. J Virol 71:6611–6618
16. Robertson ES, Lin J, Kieff E (1996) The amino-terminal domains of Epstein-Barr virus nuclear proteins 3A, 3B, and 3C interact with RBPJ (kappa). J Virol 70:3068–3074
17. Kaye KM, Devergne O, Harada JN, Izumi KM, Yalamanchili R, Kieff E, Mosialos G (1996) Tumor necrosis factor receptor associated factor 2 is a mediator of NF-kappa B acti-

vation by latent infection membrane protein 1, the Epstein-Barr virus transforming protein. Proc Natl Acad Sci USA 93:11085–11090

18. Miller WE, Mosialos G, Kieff E, Raab-Traub N (1997) Epstein-Barr virus LMP1 induction of the epidermal growth factor receptor is mediated through a TRAF signaling pathway distinct from NF-kappaB activation. J Virol 71:586–594

19. Caldwell RG, Wilson JB, Anderson SJ, Longnecker R (1998) Epstein-Barr virus LMP2A drives B cell development and survival in the absence of normal B cell receptor signals. Immunity 9:405–411

20. Sharipo A, Imreh M, Leonchiks A, Imreh S, Masucci MG (1998) A minimal glycine-alanine repeat prevents the interaction of ubiquitinated I kappaB alpha with the proteasome: a new mechanism for selective inhibition of proteolysis [see comments]. Nat Med 4:939–944

21. Kienzle N, Sculley TB, Greco S, Khanna R (1999) Cutting edge: silencing virus-specific cytotoxic T cell-mediated immune recognition by differential splicing: a novel implication of RNA processing for antigen presentation. J Immunol 162:6963–6966

22. Rooney CM, Gregory CD, Rowe M, Finerty S, Edwards C, Rupani H, Rickinson AB (1986) Endemic Burkitt's lymphoma: phenotypic analysis of tumor biopsy cells and of derived tumor cell lines. J Natl Cancer Inst 77:681–687

23. Rowe M, Khanna R, Jacob CA, Argaet V, Kelly A, Powis S, Belich M, Croom-Carter D, Lee S, Burrows SR, et al (1995) Restoration of endogenous antigen processing in Burkitt's lymphoma cells by Epstein-Barr virus latent membrane protein-1: coordinate upregulation of peptide transporters and HLA-class I antigen expression. Eur J Immunol 25(5):1374–84

24. Zeidler R, Eissner G, Meissner P, Uebel S, Tampe R, Lazis S, Hammerschmidt W (1997) Downregulation of TAP1 in B lymphocytes by cellular and Epstein-Barr virus-encoded interleukin-10. Blood 90:2390–2397

25. Drexler HG (1992) Recent results on the biology of Hodgkin and Reed-Sternberg cells. I. Biopsy material. Leuk Lymphoma 8:283–313

26. Murray PG, Constandinou CM, Crocker J, Young LS, Ambinder RF (1998) Analysis of major histocompatibility complex class I, TAP expression, and LMP2 epitope sequence in Epstein-Barr virus-positive Hodgkin's disease. Blood 92:2477–2483

27. Herbst H, Foss H-D, Samol J, Araujo I, Klotzbach H, Krause H, Agathanggelou A, Niedobitek G, Stein H (1996) Frequent expression of interleukin-10 by Epstein-Barr virus-harboring tumor cells of Hodgkin's disease. Blood 87:2918–2929

28. Hsu SM, Lin J, Xie SS, Hsu PL, Rich S (1993) Abundant expression of transforming growth factor-beta 1 and -beta 2 by Hodgkin's Reed-Sternberg cells and by reactive T lymphocytes in Hodgkin's disease. Hum Pathol 24:249–255

29. Ohshima K, Suzumiya J, Akamatu M, Takeshita M, Kikuchi M (1995) Human and viral interleukin-10 in Hodgkin's disease, and its influence on CD4+ and CD8+ T lymphocytes. Int J Cancer 62:5–10

30. Poppema S, Potters M, Visser L, van den Berg AM (1998) Immune escape mechanisms in Hodgkin's disease. Ann Oncol 9(suppl 5):S21–S24

31. Rickinson AB, Moss DJ (1997) Human cytotoxic T lymphocyte responses to Epstein-Barr virus infection. Annu Rev Immunol 15:405–431

32. Aguilar LK, Rooney CM, Heslop HE (1999) Lymphoproliferative disorders involving Epstein-Barr virus after hemopoietic stem cell transplantation. Curr Opin Oncol 11:96–101

33. Lucas KG, Burton RL, Zimmerman SE, Wang J, Cornetta KG, Robertson KA, Lee CH, Emanuel DJ (1998) Semiquantitative Epstein-Barr virus (EBV) polymerase chain reac-

tion for the determination of patients at risk for EBV-induced lymphoproliferative disease after stem cell transplantation. Blood 91:3654–3661

34. Heslop HE, Brenner MK, Rooney CM (1994) Donor T cells to treat EBV-associated lymphoma. N Eng J Med 331:679–680

35. Smith CA, Ng CYC, Heslop HE, Holladay MS, Richardson S, Turner EV, Loftin SK, Li C, Brenner MK, Rooney CM (1995) Production of genetically modified EBV-specific cytotoxic T cells for adoptive transfer to patients at high risk of EBV-associated lymphoproliferative disease. J Hematother 4:73–79

36. Rooney CM, Wimperis JZ, Brenner MK, Patterson J, Hoffbrand AV, Prentice HG (1986) Natural killer cell activity following T-cell depleted allogeneic bone marrow transplantation. Br J Haematol 62:413–420

37. Heslop HE, Brenner MK, Rooney CM, Krance RA, Roberts WM, Rochester R, Smith CA, Turner V, Sixbey J, Moen R, Boyett JM (1994) Administration of neomycin-resistance-gene-marked EBV-specific cytotoxic T lymphocytes to recipients of mismatched-related or phenotypically similar unrelated donor marrow grafts. Hum Gene Ther 5:381–397

38. Rooney CM, Smith CA, Ng CYC, Loftin SK, Sixbey JW, Gan Y-J, Srivastava D-K, Bowman LC, Krance RA, Brenner MK, Heslop HE (1998) Infusion of cytotoxic T cells for the prevention and treatment of Epstein-Barr virus-induced lymphoma in allogeneic transplant recipients. Blood 92:1549–1555

39. Rooney CM, Smith CA, Ng C, Loftin SL, Li C, Krance RA, Brenner MK, Heslop HE (1995) Use of gene-modified virus-specific T lymphocytes to control Epstein-Barr virus-related lymphoproliferation. Lancet 345:9–13

40. Gottschalk S, Ng CYC, Perez M, Brenner MK, Heslop HE, Rooney CM (1998) Mutation in EBV produces immunoblastic lymphoma unresponsive to CTL immunotherapy. Blood Suppl 1:(abstract 1316 pp321a)

41. Haque T, Amlot PL, Helling N, Thomas JA, Sweny P, Rolles K, Burroughs AK, Prentice HG, Crawford DH (1999) Reconstitution of EBV-specific T cell immunity in solid organ transplant recipients. J Immunol 160(12): 6204–6209

42. Roskrow MA, Suzuki N, Gan Y-J, Sixbey JW, Ng CYC, Kimbrough S, Hudson MM, Brenner MK, Heslop HE, Rooney CM (1998) EBV-specific cytotoxic T lymphocytes for the treatment of patients with EBV positive relapsed Hodgkin's disease. Blood 91:2925–2934

43. Nair S, Babu JS, Dunham RG, Kanda P, Burke RL, Rouse BL (1993) Induction of primary, antiviral cytotoxic and proliferative responses with antigens administered via dendritic cells. J Virol 67:4062–4069

44. Choudhury A, Toubert A, Sutaria S, Charron D, Champlin RE, Claxton DF (1998) Human leukemia-derived dendritic cells: ex-vivo development of specific antileukemic cytotoxicity. Crit Rev Immunol 18:121–131

Processing of Antigens by Dendritic Cells: Nature's Adjuvant

KAYO INABA

Summary. These are exciting times in dendritic cell biology. For example, scientists are learning to control dendritic cell development to produce large numbers of these specialized antigen-presenting cells. The dendritic cells are then used in studies of mechanism of action, e.g., antigen uptake and processing. The findings also are being exploited in clinical studies where antigen-pulsed dendritic cells are being administered to initiate immunity in humans, especially resistance to tumors. Of recent interest is the capacity of dendritic cells to process antigens from dying cells. Dendritic cells process antigens from apoptotic cells to produce MHC class I–peptide complexes, and also process either apoptotic or necrotic cells to produce MHC class II–peptide complexes. The efficacy of antigen presentation by dendritic cells can be studied with standard T-cell stimulation assays or with monoclonal antibodies that directly identify specific MHC–peptide complexes. We have used antibodies to show that dendritic cells efficiently present peptides from dying cells. Dendritic cells also express high levels of costimulatory molecules (e.g., CD40, CD86) for T-cell stimulation, and select chemokine receptors (e.g., CCR7) that guide the dendritic cells to the T-cell areas of lymphoid organs.

Key words. Dendritic cells, Antigen presentation, Tumors, Adjuvant

Many investigators would like to use autologous dendritic cells to immunize patients against clinically important antigens, such as tumor antigens (reviewed in [1,2]). This interest in dendritic cells is based on many findings. Here I would like to review some features of dendritic cells, explain why these cells are being considered for immune therapy, and then consider recent

Department of Zoology, Graduate School of Science, Kyoto University, Kitashirakawa-Oiwake-cho, Sakyo-ku, Kyoto 606-8502, Japan

observations on the capacity of dendritic cells to take up and process antigens.

Dendritic cells are unusual antigen-presenting cells [3,4]. They are potent, which means that small numbers of dendritic cells and low doses of antigen lead to strong T-cell responses. Dendritic cells can prime CD4, CD8, and NK-T cells, which means that they can initiate immunity both in culture and in vivo. Additionally, dendritic cells are physiological antigen-presenting cells, which means that they are found in the right place in vivo to pick up antigens, and they can migrate to lymphoid organs to initiate immunity [5–7].

Almost 10 years ago, we took dendritic cells from mouse spleen, or we grew dendritic cells from progenitors in bone marrow, and then we pulsed the cells with protein and microbial antigens ex vivo. The antigen-pulsed dendritic cells then were reinjected into syngeneic animals. We found that we could prime T cells to the specific antigens that were pulsed onto the dendritic cells [8]. We used major histocompatibility complex (MHC) restriction to prove that the dendritic cells were priming the mice directly. We primed CD4 helper cells in these experiments, while others found that dendritic cells would prime CD8 killer cells. Several laboratories now have reported that antigen-pulsed dendritic cells can elicit protection against tumors and infections, for example [9–14]. Thus, the topic I develop here is the loading of dendritic cells with antigen, because it is helpful to optimize the number and amount of antigens presented on the dendritic cells for optimal immune therapy.

One system that is attracting much attention is one that was pioneered by Albert et al., who showed that dendritic cells could present antigens from other cells [15]. They were studying dendritic cells from HLA-A2.1 individuals in whom strong CD8+ CTL responses could be induced to influenza antigens. Albert et al. then placed the influenza virus or the influenza matrix gene in HLA-A2.1-negative cells. If the A2.1-negative cells were made to undergo apoptosis, then the HLA-2.1-positive dendritic cells phagocytosed the apoptotic cells and presented the peptides on MHC class I [15,16]. Macrophages in contrast destroyed phagocytosed dying cells but did not form detectable MHC–peptide complexes [15,16]. We decided to pursue this processing of other cells with mouse dendritic cells. First, I explain how we obtained large numbers of mouse dendritic cells, and also explain the term "immature dendritic cell," because the immature stage is the one that phagocytoses dying cells.

We grew dendritic cells from marrow progenitors with granulocyte-macrophage colony-stimulating factor (GM-CSF) [17]. The progenitors grow in distinctive aggregates. In the aggregates, the dendritic cells are immature

[17,18]; in other words, they do not yet have potent T-cell-stimulating activity. However, the immature dendritic cells are endocytic and phagocytose dying cells and microorganisms [16,19–21]. Also, the immature cells have abundant MHC II products within lysosomal compartments, often called MIICs [18,19,22,23]. When the dendritic cells are made to mature, they express large amounts of T-cell costimulatory molecules, and the MHC II moves to the cell surface where the molecules are long lived [18,24].

To follow the antigen-presenting activity of bone marrow dendritic cells, we fed the immature cells a protein antigen, hen egg lysozyme (HEL). We then measured the response of HEL-specific, TCR-transgenic T cells using tritiated thymidine uptake. If the dendritic cells are not given a maturation stimulus such as lipopolysaccharide (LPS), they do not present antigen to the T cells; however, if they are given both HEL antigen and LPS, then antigen presentation is very strong. Even 150 dendritic cells induce a clear response in 200000 naïve T cells. Therefore, dendritic cell maturation is an important control point in the initiation of immunity. Maturation stimuli like LPS or CD40L or TNF allow dendritic cells to express high levels of costimulators and to efficiently present antigens.

We now return to the processing of whole cells, because this may be critical in the development of dendritic cell-based immunotherapy. We used a monoclonal antibody called Y-Ae, which was provided by Dr. Charles Janeway, that is like a T-cell receptor because it recognizes a complex formed by an MHC product, I-A^b presenting a specific peptide from I-E [25]. Our experiment was to take dendritic cells from C57BL/6 I-A^b mice and feed them BALB/C B cells that expressed the I-E peptide. If the C57BL/6 dendritic cells processed the peptide out of the B cells, we would see the expression of Y-Ae on the dendritic cells. We added the dying B cells to phagocytic immature dendritic cells, and we also induced the dendritic cells to mature, usually by dislodging the cells and transferring them to a new culture vessel. We looked for Y-Ae on the dendritic cells after an overnight culture of about 20 h. Y-Ae was expressed if we fed the dendritic cells preprocessed I-E peptide, but we also saw the Y-Ae signal if we fed the dendritic cells I-E-positive but not I-E-negative B blasts [19]. No Y-Ae formed if we separated the dendritic cells and B cells by a filter, or if we fed B blasts to dendritic cells that were already mature.

Shannon Turley (in Ira Mellman's lab) then observed the cultures by immunofluorescence confocal microscopy. We found that B-cell fragments were phagocytosed by the immature dendritic cells [19]. Most of the B-cell fragments were delivered to the MHC II compartments of the immature dendritic cells. In other experiments, we gave the cells ammonium chloride to reversibly block endosomal and lysosomal processing. We then purified the dendritic cells away from the remaining B cells and removed the ammonium chloride. The B cells were quickly digested, while the MHC–peptide complexes recognized by the Y-Ae antibody appeared on the dendritic cell surface. The purified fixed dendritic cells also efficiently presented antigen to a T-cell hybridoma that was specific for the same MHC–peptide complex as Y-Ae.

Kinetic studies were also carried out on the formation of the Y-Ae MHC–peptide complex. We measured mean fluorescence index of the Y-Ae signal following administration of graded doses of I-E peptide or I-E-positive B blasts. A dose of ten blasts per dendritic cell gave the same Y-Ae signal as 1 micromolar preprocessed peptide. We also were able to quantitate the amount of I-E protein in the B cells, using a polyclonal anti-I-E antibody that was kindly provided by Dr. Ron Germain. This antibody detects all forms of cellular I-E, including molecules that are not dimerized or associated with invariant chain. What was striking was the efficiency with which I-E molecules within phagocytosed B cells were processed to form MHC–peptide complexes; 1 micromolar of preprocessed peptide gave a similar signal to less than 1 nanomolar of protein in B cells (Fig. 1). So dying cells, if delivered to the dendritic cell at the right stage of development, are more than a thousand times better than peptide to make MHC–peptide complex [19]. Therefore, dendritic cells must be very efficient at taking up dying cells and processing them to make MHC–peptide complexes. Dendritic cells could process either apoptotic or necrotic cells onto MHC class II, but for class I presentation, apoptotic cells seem to be necessary.

We also have used another antibody called C4H3 to another MHC peptide complex formed by I-A^k presenting a peptide from HEL [26]. This antibody

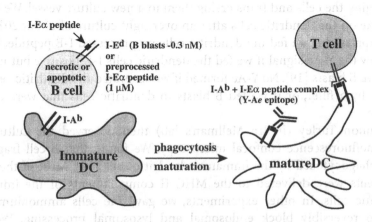

FIG. 1. Presentation of phagocytosed cellular fragments by dendritic cells

also was generously provided by Dr. Ron Germain and Caetano Reis e Sousa. We found that the HEL was delivered to MIICs and converted to MHC–peptide complexes. Therefore the use of monoclonal antibodies to MHC–peptide complexes is proving to be a direct way to demonstrate the efficient antigen-processing and presentation capacities of maturing dendritic cells.

In summary, one can list three types of features of dendritic cells as antigen presenting cells. First, they are very efficient at capturing antigens and forming high levels of MHC peptide, but this efficiency is evident during their maturation in response to microbial and inflammatory stimuli. Second, dendritic cells express a number of membrane costimulatory molecules and can produce stimulatory cytokines like IL-12. Finally, dendritic cells have a number of properties in vivo that are useful for initiating immunity. Dendritic cells undergo many important changes when stimulated to mature, for example, upon exposure to lipopolysaccharide or CD40 ligand (Fig. 2). The maturing dendritic cells form high levels of MHC–peptide complexes and high levels of costimulatory molecules. Other laboratories have shown that maturing dendritic cells also express CCR7 [27–29]. This chemokine receptor is thought to be important in directing the mature dendritic cell to the T-cell area [27,30] where it is known that they initiate immune responses [7,31,32].

Dendritic cells can be very efficient at capturing antigens from other cells (Fig. 3). I have discussed processing of apoptotic and necrotic cells, but there are other pathways, such as processing of exosomes and immune complexes. This capture of antigens from other cells may be important in many clinical conditions, such as the indirect pathway of graft rejection or cross priming, the presentation of dying infected cells, and self-tolerance.

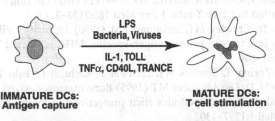

LPS
Bacteria, Viruses
IL-1, TOLL
TNFα, CD40L, TRANCE

IMMATURE DCs:
Antigen capture

apoptotic cells
necrotic cells
microbes
FITC-Dextran
soluble proteins

MATURE DCs:
T cell stimulation

costimulatory molecules
(CD40, 54, 58, 80, 86)
high MHC-peptide complexes
resist to IL-10, produce IL-12
CCR7, CCR4, CXCR4

FIG. 2. Maturation of dendritic cells

Cross presentation :cross priming, cross tolerance

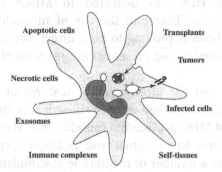

Fig. 3. Processing of antigens from other cells by dendritic cells

References

1. Fernandez N, Duffour M-T, Perricaudet M, Lotze MT, Tursz T, Zitvogel L (1998) Active specific T-cell-based immunotherapy for cancer: nucleic acids, peptides, whole native proteins, recombinant viruses, with dendritic cell adjuvants or whole tumor cell-based vaccines. Principles and future prospects. Cytokines Cell Mol Ther 4:53–65
2. Schuler G, Steinman RM (1997) Dendritic cells as adjuvants for immune-mediated resistance to tumors. J Exp Med 186:1183–1187
3. Hart DNJ (1997) Dendritic cells: unique leukocyte populations which control the primary immune response. Blood 90:3245–3287
4. Banchereau J, Steinman RM (1998) Dendritic cells and the control of immunity. Nature (Lond) 392:245–252
5. Larsen CP, Steinman RM, Witmer-Pack M, Hankins DF, Morris PJ, Austyn JM (1990) Migration and maturation of Langerhans cells in skin transplants and explants. J Exp Med 172:1483–1493
6. Austyn JM (1996) New insights into the mobilization and phagocytic activity of dendritic cells. J Exp Med 183:1287–1292
7. Ingulli E, Mondino A, Khoruts A, Jenkins MK (1997) In vivo detection of dendritic cell antigen presentation to CD4+ T cells. J Exp Med 185:2133–2141
8. Inaba K, Metlay JP, Crowley MT, Steinman RM (1990) Dendritic cells pulsed with protein antigens in vitro can prime antigen-specific, MHC-restricted T cells in situ. J Exp Med 172:631–640
9. Mayordomo JI, Zorina T, Storkus WJ, Zitvogel L, Celluzzi C, Falo LD, Melief CJ, Ilstad ST, Kast WM, DeLeo AB, Lotze MT (1995) Bone marrow-derived dendritic cells pulsed with synthetic tumour peptides elicit protective and therapeutic antitumour immunity. Nat Med 1:1297–1302
10. Specht JM, Wang G, Do MT, Lam JS, Royal RE, Reeves ME, Rosenberg SA, Hwu P (1997) Dendritic cells retrovirally transduced with a model tumor antigen gene are therapeutically effective against established pulmonary metastases. J Exp Med 186:1213–1221
11. Song W, Kong H, Carpenter H, Torii H, Granstein R, Rafii S, Moore MAS, Crystal RG (1997) Dendritic cells genetically modified with an adenovirus vector encoding the

cDNA for a model tumor antigen induce protective and therapeutic antitumor immunity. J Exp Med 186:1247–1256

12. Nestle FO, Alijagic S, Gilliet M, Sun Y, Grabbe S, Dummer R, Burg G, Schadendorf D (1998) Vaccination of melanoma patients with peptide- or tumor lysate-pulsed dendritic cells. Nat Med 4:328–332

13. Ludewig B, Ehl S, Karrer U, Odermatt B, Hengartner H, Zinkernagel RM (1998) Dendritic cells efficiently induce protective antiviral immunity. J Virol 272:3812–3818

14. Su H, Messer R, Whitmire W, Fischer E, Portis JC, Caldwell HD (1998) Vaccination against chlamydial genital tract infection following immunization with dendritic cells pulsed ex vivo with non-viable chlamydiae. J Exp Med 188:809–818

15. Albert ML, Sauter B, Bhardwaj N (1998) Dendritic cells acquire antigen from apoptotic cells and induce class I-restricted CTLs. Nature (Lond) 392:86–89

16. Albert ML, Pearce SFA, Francisco LM, Sauter B, Roy P, Silverstein RL, Bhardwaj N (1998) Immature dendritic cells phagocytose apoptotic cells via $\alpha_v\beta_5$ and CD36, and cross-present antigens to cytotoxic T lymphocytes. J Exp Med 188:1359–1368

17. Inaba K, Inaba M, Romani N, Aya H, Deguchi M, Ikehara S, Muramatsu S, Steinman RM (1992) Generation of large numbers of dendritic cells from mouse bone marrow cultures supplemented with granulocyte/macrophage colony-stimulating factor. J Exp Med 176:1693–1702

18. Pierre P, Turley SJ, Gatti E, Hull M, Meltzer J, Mirza A, Inaba K, Steinman RM, Mellman I (1997) Developmental regulation of MHC class II transport in mouse dendritic cells. Nature (Lond) 388:787–792

19. Inaba K, Turley S, Yamaide F, Iyoda T, Mahnke K, Inaba M, Pack M, Subklewe M, Sauter B, Sheff D, Albert M, Bhardwaj N, Mellman I, Steinman RM (1998) Efficient presentation of phagocytosed cellular fragments on the MHC class II products of dendritic cells. J Exp Med 188:2163–2173

20. Inaba K, Inaba M, Naito M, Steinman RM (1993) Dendritic cell progenitors phagocytose particulates, including Bacillus Calmette-Guerin organisms, and sensitize mice to mycobacterial antigens in vivo. J Exp Med 178:479–488

21. Reis e Sousa C, Stahl PD, Austyn JM (1993) Phagocytosis of antigens by Langerhans cells in vitro. J Exp Med 178:509–519

22. Sallusto F, Cella M, Danieli C, Lanzavecchia A (1995) Dendritic cells use macropinocytosis and the mannose receptor to concentrate antigen in the major histocompatibility class II compartment. Downregulation by cytokines and bacterial products. J Exp Med 182:389–400

23. Cella M, Sallusto F, Lanzavecchia A (1997) Origin, maturation and antigen presenting function of dendritic cells. Curr Opin Immunol 9:10–16

24. Cella M, Engering A, Pinet V, Pieters J, Lanzavecchia A (1997) Inflammatory stimuli induce accumulation of MHC class II complexes on dendritic cells. Nature (Lond) 388:782–787

25. Rudensky AY, Rath S, Preston-Hurlburt P, Murphy DB, Janeway CA Jr (1991) On the complexity of self. Nature (Lond) 353:660–662

26. Zhong G, Reis e Sousa C, Germain RN (1997) Antigen-unspecific B cells and lymphoid dendritic cells both show extensive surface expression of processed antigen-major histocompatibility complex class II complexes after soluble protein exposure in vivo or in vitro. J Exp Med 186:673–682

27. Dieu M-C, Vanbervliet B, Vicari A, Bridon J-M, Oldham E, Ait-Yahia S, Briere F, Zlotnik A, Lebecque S, Caux C (1998) Selective recruitment of immature and mature dendritic cells by distinct chemokines expressed in different anatomic sites. J Exp Med 188:373–386

28. Yanagihara S, Komura E, Nagafune J, Watarai H, Yamaguchi Y (1998) EB11/CCR7 is a new member of dendritic cell chemokine receptor that is upregulated upon maturation. J Immunol 161:3096–3102

29. Sallusto F, Schaerli P, Loetscher P, Schaniel C, Lenig D, Mackay CR, Qin S, Lanzavecchia A (1998) Rapid and coordinated switch in chemokine receptor expression during dendritic cell maturation. Eur J Immunol 28:2760–2769

30. Ngo VN, Tang HL, Cyster JG (1998) Epstein-Barr virus-induced molecule 1 ligand chemokine is expressed by dendritic cells in lymphoid tissues and strongly attracts naive T cells and activated B cells. J Exp Med 188:181–191

31. Kudo S, Matsuno K, Ezaki T, Ogawa M (1997) A novel migration pathway for rat dendritic cells from the blood: hepatic sinusoids-lymph translocation. J Exp Med 185:777–784

32. Luther SA, Gulbranson-Judge A, Acha-Orbea H, Maclennan ICM (1997) Viral superantigen drives extrafollicular and follicular B differentiation leading to virus-specific antibody production. J Exp Med 185:551–562

Immunotherapy of Melanoma Using Dendritic Cells

FRANK O. NESTLE

Summary. Melanoma is a malignant skin tumor of melanocytic origin. No cure is currently available in advanced stages with distant metastasis. Recent progress in the understanding of mechanisms of immune activation and immune escape during an antimelanoma-specific immune response has resulted in new concepts for immunotherapeutic intervention in this disease. In a clinical pilot trial, 30 metastatic melanoma patients were vaccinated with peptide- and/or tumor lysate-pulsed cells. All patients developed a strong delayed-type hypersensitivity (DTH) reaction to the tracer molecule KLH. Peptide-specific immune response could be detected by DTH to peptide-pulsed dendritic cells (DC) and was correlated to response to therapy. DTH reactivity to peptide alone was detected in 6 patients. Clinical responses were induced in 27% (8/30) of the patients including 3 CR (complete remissions) and 5 PR (partial remissions). Immune escape mechanisms were evident at various levels of antigen presentation, including defects in expression of proteasomal antigens, TAP deficiency, melanoma antigen loss variants, and absent expression of relevant HLA surface molecules. DC vaccination for induction of an antitumor response in melanoma patients is safe and promising. However, we believe that aside from the optimal strategy for the induction of an immune response, factors such as a tumor immune escape mechanisms also have to be considered as limitations for therapy.

Key words. Melanoma, Dendritic cells, Immunotherapy, Antigen-presenting cells, Immunointervention

Department of Dermatology, University of Zurich Medical School, Gloriastrasse 31, 8091 Zurich, Switzerland

Introduction

Dendritic cells (DC) are highly specialized antigen-presenting cells (APC) for the activation of antigen-specific effector T cells [1]. In the past years an enormous increase in our understanding of DC biology has opened new ways for the application of these cells for immunotherapy of cancer [2]. We discuss current issues in the generation and application of DC, use of a surrogate helper antigen, various methods to detect antigen-specific immune responses, and some of our clinical data relating to vaccination of stage IV melanoma patients with antigen-pulsed DC. Finally, we propose reasons for lack of response to DC therapy.

Generation and Application of DC

The possibility to differentiate DC in the presence of certain cytokines such as IL-4 and granulocyte-macrophage colony-stimulating factor (GM-CSF) from plastic adherent monocyte precursors [3,4] opened the field of DC biology for possible therapeutic applications. At the same time definitions and criteria for DC became unclear. Is a plastic adherent cell generated in serum-free medium in the presence of IL-4 and GM-CSF during 7 days from blood monocytes a "real DC"? How stable is the phenotype of this cell and what kind of maturity is represented by this cell type? Do we need additional maturation stimuli to create a stable phenotype and more potent T-cell stimulatory capacity? What is the migratory capacity of the various DC types? Is an immature or mature DC better suited for induction of a tumor antigen-specific T-cell response in vivo?

These types of questions are central issues to current DC therapy. Because the dichotomy of immature and mature DC is not easily transferable from men to mice, the necessary questions are not easy to address in rodents. Therefore only a wide variety of clinical pilot studies with carefully designed and performed readout systems will give us an answer about the optimal DC preparation to use. At the time when we started to use DC for clinical application, the only technique available was generation in fetal calf serum (FCS) containing medium in the presence of IL-4 and GM-CSF. These FCS-DC carry high levels of major histocompatibility complex (MHC) molecules and costimulatory molecules on their surface and are potent stimulators of T cells. They differ form their counterparts generated in serum-free medium by higher expression of costimulatory molecules and better T-cell stimulatory capacity. Currently we prefer to use DC generated under serum-free conditions. Appropriate maturation stimuli would be lipopolysaccharide (LPS) or CD40L, but neither reagent is available in sufficient quantity for clinical use. We started to use monocyte-conditioned supernatant to mature DC, a good tool to induce maturation [5,6]. However, the cocktail has the disadvantage that the relevant maturation-inducing cytokines may change from preparation to preparation. Therefore we switched to a defined cytokine cocktail [IL-1, IL-6, tumor necrosisfactor-α (TNF-α) or prostaglandin E_2 (PGE$_2$)] as a maturation stimulus [7]. This cocktail provides a reliable maturation of DC and a stable phenotype with potent T-cell stimulatory capacity. We prefer for quality control the use of markers relevant to the function of DC such as CD80, CD86, CD40, and HLA-DR as well as surrogate markers for mature DC such as CD83.

In our view an even more important point than the method to generate DC is the way they are injected into the patient. Several possible injection modes may be envisaged, such as intravenous, intradermal, subcutaneous, or intranodal. Intravenous injection is currently used by several groups. Concerns may be raised that injected DC will be first trapped in the lung, leading to disruption of DC-antigen complexes. Free antigen may disseminate in the body and may be presented in the wrong immunological context, leading to induction of tolerance instead of immune activation. Furthermore, large numbers of DC are currently used for intravenous application, which necessitates leukapheresis to obtain sufficient DC precursors, a strenuous and rather unpleasant procedure for the patient. Intradermal injection seems to be an appropriate method if we extrapolate data obtained from animal models. Intradermally injected DC are supposed to migrate through the afferent lymphatics to a tissue-draining lymph node. No studies have yet convincingly demonstrated or tested the migratory capacity of the various human DC preparations in vivo. Migratory capacity and also the number of DC that finally reach the lymph node are critical issues for future studies.

Because convincing data about the migratory capacity of human DC were lacking when we started our study, we decided to inject our DC preparation directly under ultrasound control in a normal-appearing inguinal lymph node. As discussed, the lymph node is the appropriate meeting point for a DC-T-cell cross-talk and is, at least from a theoretical point of view, the preferred injection site for antigen-loaded DC. DC are injected at 1×10^6 cells in 500 μl of PBS. Injection is controlled by ultrasound. Increase of the hypoechogen area of a lymph node corresponds to the injected DC-containing fluid. Injection efficiency is very high and reproducible (M. Hauser and F.O. Nestle, manuscript submitted).

KLH As a Surrogate Helper Antigen

Peptide pulsing of DC will probably generate only a few specific peptide-MHC complexes by competing with low-affinity binding self-peptides and labeling of some "empty" MHC class I surface molecules. Chances are low that under these circumstances antigen-specific cytotoxic T cells (CTL) will be generated in the absence of additional signals. Several animal models have shown that "weak antigens" such as tumor antigens need help to generate effective antigen-specific CTL. Help may derive from an appropriate cytokine environment or the presence of specialized helper T cells. We reasoned that the introduction of a surrogate helper antigen would generate appropriate help signals. We choose keyhole limpet hemocyanin (KLH), a snail-derived neoantigen with potent immunostimulatory capacity. This antigen would allow us to (i) control for the induction of a CD4-positive helper-T-cell response and (ii) introduce a potent helper-T-cell response at the site of T-cell priming (activation), i.e., in the respective injected lymph node. KLH was delivered to DC through the exogenous pathway of antigen presentation and indeed induced after intranodal injection a very strong KLH-specific memory response in vivo (M. Gilliet and F.O. Nestle, manuscript submitted). Abundant KLH-specific memory T cells in the injected lymph node will contribute to the maturation and induction of IL-12 production by DC through CD40/CD40L signaling and will also lead to the generation of so-called conditioned or superactivated DC that are able to activate killer T cells in the absence of help [8,9].

Detection of an Antigen-Specific Immune Response

Detection of an antigen-specific immune response is an important surrogate marker to control for an effective vaccination strategy even though the correlation with clinical response is the most important issue. The classic way to detect CTL activity is the measurement of lytic activity against ^{51}Cr-labeled target cells. Because precursor frequencies are low, two to three in vitro restimulations are necessary to reach a CTL frequency detectable by cytotoxicity assays. These techniques are time consuming and may introduce an artifact that limits quantitative analysis. Still, this method is the gold standard because it measures lytic activity of effector cells, the same functional activity we would like to induce in vivo.

Recently introduced methods rely on measurement of release of cytokines by CTL after contact with antigen. Cytokines may be measured by an ELISA method (ELISPOT) or quantified by intracellular cytokine staining and detection by flow cytometry. Problems with the ELISPOT method are the low reproducibility of spot counting by various individuals, which may be overcome with the introduction of sophisticated imaging systems. Another drawback is that detection of cytokine release does not necessarily correlate with the cytolytic activity of a given cell. We perform an additional method, which is delayed-type hypersensitivity (DTH) tesing for peptide-specific immune responses. Peptide alone or DC pulsed with peptide are injected intradermally, and induration as well as erythema is read after 48h. We obtained very good correlation between a positive DTH response and clinical response in all our patients. This technique is also of importance for both the patient and the physician in charge because they can observe immediately if there is a suggestion of immunoreactivity. Peptide DTH testing has been successfully used in various mouse models [10]. Furthermore, we could demonstrate the recruitment of peptide-specific T cells to the injection site in humans [11].

A method recently much talked about is the use of HLA–peptide tetrameric complexes [12–14]. A biotinylation site is introduced in in vitro synthesized MHC–peptide complexes. After biotinylation, tetramer formation is induced by adding fluorescent-labeled streptavidin. These complexes bind to antigen-specific T cells, which may be identified on a per-cell basis by flow cytometry. Problems with this method are low sensitivity, which often necessitates in vitro restimulation of antigen-specific T cells. Furthermore, tetramers may not be able to detect low- or intermediate-affinity T cells, which are of importance especially in the context of vaccination against self-antigens (differentiation antigens in melanoma).

Side Effects

In more than 30 advanced melanoma patients treated with peptide- or lysate-pulsed DC, no significant side effects occurred (M. Gilliet and F.O. Nestle, manuscript submitted). In a few patients, slight fever or painful lymph nodes occurred after vaccination. Because we use peptide antigens derived from melanocytic differentiation antigens, autoimmune-like reactions may occur. Progressive vitiligo as well as depigmentation of melanocytic nevi may occur during vaccination. No serious destructive autoimmunity was observed except induction of antithyroid-receptor autoantibodies, antinuclear antibodies, and rheumatoid factor in a few patients.

Clinical Results

Even though the induction of an antigen-specific immune response is an important readout to control for successful vaccination, our final goal is the regression of tumor lesions. Strong criteria have to be applied to responding patients to evaluate success of therapy. Of special interest are the objective response rate (complete and partial responses) and the duration of responses, as well as the body site where metastases regress. In more than 30 treated patients, objective response rates are in agreement with earlier published data [11; submitted], indicating that now larger multicenter studies and comparison to established treatment regimens for advanced melanoma should be considered to provide insight into the real clinical efficacy of DC vaccination.

There are several interesting clinical observations, which we can discuss only briefly here. There are multiple clinical complete responses with durations up to 2 years, indicating a long-lasting antitumor effect. Unpublished preliminary data support the idea that repeated vaccination may be necessary to sustain response to therapy. Tumor regression occurs at various tissue sites including lung, pancreas, lymph nodes, and skin. There is a very good correlation between response to therapy and immunological response. Repetitive vaccination may be beneficial to patients, and response to vaccination during repetitive vaccination correlates with immunological response. In all except one responder of our earlier published study, induction of peptide-specific CTL was demonstrated in the classical 4-h chromium release assay. One dogma of immunotherapy was that only patients with low tumor burden respond to vaccination therapy. We have indications that even patients with a large tumor burden may respond to DC vaccination and that response depends to a large extend on the expression of target antigens on tumor cells and not on tumor burden.

Reasons for Lack of Response to Therapy

While there is increasing optimism and accumulating evidence from various studies that DC vaccination may be indeed a very potent way to induce antigen-specific immune responses with consecutive clinical responses in cancer patients, much less is known about reasons why patients do *not* respond to therapy. Much can be learned from insights into mechanisms of immune escape, which should provide us with new concepts to increase vaccination efficiency. A prerequisite for effective killing is the expression of the respective peptide–MHC target on the surface of a melanoma cell, which may be recognized by a killer T cell. Thus, the best killer T cell induced by a "superactivated" DC will not be able to kill if the respective target is not present on the tumor cell surface.

A tumor antigen must go a long way until it is displayed on the surface of a tumor cells. It has to be transcribed and translated into a protein. Most of the proteins will pass the proteasome and are cleaved into small peptides. These peptides have to be transported into the endoplasmaic reticulum, where loading on MHC molecules takes place. Finally, the MHC–peptide complex travels through the cytoplasm and has be to displayed on the cell surface to be recognized by a killer T cell.

We investigated the expression of various proteins necessary for antigen processing and display in freshly excised metastases in situ. In all nonresponder patients to DC vaccination investigated so far, various losses and defects of molecules involves in antigen processing and display were observed (M. Gilliet and F.O. Nestle, manuscript submitted). We conclude that, even with the best vaccination regimens, a clinical response may not be observed in a very high percentage of patients with advanced melanoma and that the prescreening for such immune escape phenomena may increase our changes for successful vaccination therapy.

Acknowledgment. Information provided in this paper has been presented and published in modified form at the 5th International Symposium on Dendritic Cells in Fundamental and Clinical Immunology, Pittsburgh, PA, USA. Grants from the Cancer League Zürich, the Swiss Cancer League, to F.O.N. are gratefully acknowledged.

References

1. Banchereau J, Steinman RM (1998) Dendritic cells and the control of immunity. Nature 392:245–252
2. Nestle FO, Burg G, Dummer R (1999) New perspectives on immunobiology and immunotherapy of melanoma. Immunol Today 20:5–7
3. Romani N, Gruner S, Kämpgen E, Lenz A, Trockenbacker B, Konwalinka G, Fritsch PO, Steinman RM, Schuler G (1994) Proliferating dendritic cell progenitors in human blood. J Exp Med 180:83–93
4. Sallusto F, Lanzavecchia A (1994) Efficient presentation of soluble antigen by cultured human dendritic cells is maintained by granulocyte/macrophage colony-stimulating factor plus interleukin-4 and downregulated by tumor necrosis factor-alpha. J Exp Med 179:1109–1118
5. Bender A, Sapp M, Schuler G, Steinman RM, Bhardwaj N (1996) Improved methods for the generation of dendritic cells from nonproliferating progenitors in human blood. J Immunol Methods 196:121–135
6. Romani N, Reider D, Heuer M, Ebner S, Kampgen E, Eibl B, Niederwieser D, Schuler G. (1996) Generation of mature dendritic cells from human blood. An improved method with special regard to clinical applicability. J Immunol Methods 196:137–151
7. Jonuleit H, Kuhn U, Muller G, Steinbrink K, Paragnik L, Schmitt E, Knop J, Enk AH (1997) Pro-inflammatory cytokines and prostaglandins induce maturation of potent immunostimulatory dendritic cells under fetal calf serum-free conditions. Eur J Immunol 27(12):3135–3142
8. Lanzavechia A (1998) License to kill. Nature 393:413–414
9. Ridge JP, di Rosa F, Matzinger P (1998) A conditioned dendritic cell can be a temporal bridge between a CD4+ T-helper and a T-killer cell. Nature 393:474–477
10. Puccetti P, Bianchi R, Fioretti MC, Ayroldi E, Uyttenhove C, Peel Av, Boon T, Grohman U (1994) Use of a skin test to determine tumor-specific CD8+ T cell reactivity. Eur J Immunol 24:1446–1452

11. Nestle FO, Alijagic S, Gilliet M, Sun Y, Grabbe S, Dummer R, Burg G, Schadendorf D (1998) Vaccination of melanoma patients with peptide- or tumor lysate pulsed dendritic cells. Nat Med 4:328–332
12. Altman JD, Moss PAH, Goulder PJR, Barouch DH, McHeyzer Williams MG, Bell JI, McMichael AJ, Davis MM (1996) Phenotypic analysis of antigen-specific T lymphocytes. Science 274:94–96
13. Dunbar PR, Ogg GS, Chen J, Rust N, van der Bruggen P, Cerundolo V (1998) Direct isolation, phenotyping and cloning of low-frequency antigen-specific cytotoxic T lymphocytes from peripheral blood. Curr Biol 8:413–416
14. Ogg GS, McMicheal AJ (1998) HLA-peptide tetrameric complexes. Curr Opin Immunol 10:393–396

Immunotherapy of Melanoma Using T-Cell-Defined Antigens

YUTAKA KAWAKAMI

Summary. The results obtained from the previous immunotherapies suggest that T cells play an important role in the in vivo rejection of melanoma. Using various methods, we have identified melanoma antigens recognized by autologous T cells. These antigens are classified as (1) tissue- (melanocyte-) specific melanosomal proteins, (2) cancer-testis antigens (proteins expressed in normal testis and various cancers, and (3) tumor-specific peptides derived from mutations in tumor cells, among others. A variety of mechanisms in generating T-cell epitopes on tumor cells was discovered. Using the MART-1 and gp100 antigens, new immunotherapies including immunization with peptides, recombinant viruses, plasmid DNAs, dendritic cells pulsed with peptides as well as adoptive transfer of cytotoxic T lymphocytes (CTL) generated in vitro by stimulation with antigenic peptides, have been developed, and phase I clinical trials have been performed in the Surgery Branch of the National Cancer Institute, USA. The immunization with the gp100$_{209(210M)}$ peptide that was modified to have high HLA-A2-binding affinity, along with incomplete Freund's adjuvant and interleukin-2, resulted in 42% response rate in patients with melanoma. These immunotherapies need further improvement based on the mechanisms of tumor escape.

Key words. Tumor antigens, Melanoma, Peptides, Dendritic cells, Immunotherapy

Introduction

Treatment of melanoma patients with a high dose of IL-2 in the Surgery Branch of the National Cancer Institute (NCI) resulted in 15% (27 of 182 patients) response rate [1]. In the subsequent trial of the adoptive transfer of

Division of Cellular Signaling, Institute for Advanced Medical Research, Keio University School of Medicine, 35 Shinanomachi, Shinjuku-ku, Tokyo 160-8582, Japan

cultured tumor-infiltrating T lymphocytes (TIL) along with IL-2, 34% (29 of 86) objective response was observed [2]. The accumulation of injected T cells at tumor sites and the autologous tumor recognition by the T cells were found to be important for the tumor regression. Infiltration of T cells and macrophages was observed in the biopsied tumors that were regressing after the treatment. These results suggest that T cells play an important role in the in vivo rejection of melanoma. Therefore, we attempted to identify antigens recognized by these T cells to understand the mechanisms for tumor recognition by autologous T cells as well as to develop immunotherapies for patients with melanoma [3,4].

Identification of Melanoma Antigens Recognized by T Cells

Melanoma-reactive CD8+ or CD4+ T cells that were restricted by MHC class I or class II molecules, respectively, were induced from peripheral blood lymphocytes (PBL) or TIL from melanoma patients. These T cells responded to only autologous melanoma cells (autologous melanoma specific), melanoma and cultured melanocytes (melanocyte specific), melanoma cells and other cancer cells (cancer reactive), and only melanoma cells (melanoma specific) (Table 1). Melanocyte-specific CD8+ cytotoxic T lymphocytes (CTL) were most frequently established from TIL in the Surgery Branch, NCI. In contrast, melanoma-reactive CD4+ T cells were found to frequently recognize unique antigens on autologous melanoma.

With these melanoma-reactive T cells, we attempted to isolate melanoma antigens using a variety of methods (Table 2). Most of the antigens were

TABLE 1. Specificity of melanoma-reactive T cells

Specificity	Autologous tumor-specific T cells	Tissue-specific shared Ag-reactive T cells	Cancer-specific shared Ag-reactive T cells	Melanoma-specific shared Ag-reactive T cells
Autologous melanoma cells	+	+	+	+
Allogeneic melanoma cells	−	+	+	+
Other cancer cells	−	−	+	−
Cultured melanocytes	−	+	−	−

TABLE 2. Methods for identification of melanoma antigens recognized by T cells

I. Isolation of antigens using melanoma-reactive T cells
   1. Isolation of MHC-binding peptides
      a. Identification and sequencing of HPLC fractionated peptides eluted from MHC
         molecules on melanoma cells
      b. Isolation of peptides from random peptide libraries
   2. Isolation of genes encoding antigens
      a. Screening melanoma cDNA libraries by testing T-cell recognition of cells
         transfected with the cDNA library
II. Evaluation of candidate proteins
   1. Isolation of candidate molecules using various methods (e.g., SEREX, RDA,
      SAGE)
   2. Evaluation of the recognition of candidate proteins by melanoma-reactive T cells
   3. Induction of melanoma-reactive T cells by stimulation with candidate proteins

Candidate proteins include tissue (melanocytes, testis) specific proteins, overexpressed cancer proteins, tumor proteins recognized by antibody, etc.

isolated using the cDNA expression cloning technique with melanoma-reactive CD8+ T cells [3,4]. Cells transfected with melanoma cDNA libraries were screened by measuring cytokine release from melanoma-reactive T cells, and cDNAs encoding melanoma antigens were isolated. T-cell epitopes were subsequently identified by testing peptides that were synthesized on the basis of MHC allele-specific peptide-binding motifs. A number of melanoma antigens recognized by T cells have been isolated. Representative melanoma antigens are (1) tissue- (melanocyte-) specific melanosomal proteins, (2) cancer-testis antigens (proteins expressed in testis and a variety of cancers), and (3) tumor-specific peptides derived from mutations in tumor cells (Table 3).

These antigens have recently been found to be also recognized by IgG antibodies in the sera of patients [5]. Thus, candidate proteins for melanoma antigens recognized by T cells may be tissue-specific (melanocytes, testis) proteins, overexpressed cancer proteins, tumor proteins recognized by antibodies, and so on. To isolate new melanoma antigens, we are attempting to induce these tumor-reactive T cells against candidate proteins. Tissue-specific (melanocytes, testis) proteins and overexpressed cancer proteins can be isolated by cDNA subtraction methods. Using RDA (representational difference analysis) and SAGE (serial analysis of gene expression), we have confirmed that known melanoma antigens including gp100 could be isolated by these techniques. Tumor proteins recognized by antibodies can be isolated by the SEREX (serological analysis of autologous tumor antigens by recombinant cDNA expression cloning) method [6].

TABLE 3. Human melanoma antigens recognized by T cells

Antigen	Ag-presenting MHC	Epitope
A. Possibly autoreactive antigens: melanosomal proteins		
gp100	HLA-A2	KTWGQYWQV[a,b]
	HLA-A2	AMLGTHTMEV
	HLA-A2	MLGTHTMEV
	HLA-A2	ITDQVPFSV[a,b]
	HLA-A2	YLEPGPVTA[a,b]
	HLA-A2	LLDGTATLRL
	HLA-A2	VLYRYGSFSV
	HLA-A2	SLADTNSLAV
	HLA-A2	RLMKQDFSV
	HLA-A2	RLPRIFCSC
	HLA-A3	ALLAVGATK
	HLA-A3	LIYRRRLMK
	HLA-A24	VYFFLPDHL[c]
MART-1/Melan-A	HLA-A2	AAGIGILTV[a,b]
	HLA-A2	EAAGIGILTV[a,b]
	HLA-A2	ILTVILGVL
	HLA-B45	AEEAAGIGIL
	HLA-B45	AEEAAGIGILT
TRP1 (gp75)	HLA-A31	MSLQRQFLR[d]
TRP2	HLA-A2	SVYDFFVWL
	HLA-A31	LLGPGRPYR
	HLA-A33	LLGPGRPYR
Tyrosinase	HLA-A1	DAEKCDKTDEY[b]
	HLA-A1	SSDYVIPIGTY
	HLA-A2	MLLAVLYCL[a,b]
	HLA-A2	YMDGTMSQV[a,b]
	HLA-A24	AFLPWHRLF
	HLA-B44	SEIWRDIDF
	HLA-DR4	QNILLSNAPLGPQFP
	HLA-DR4	SYLQDSDPDSFQD
B. Tumor-specific shared antigens		
a. Proteins expressed in testis and a variety of cancers		
MAGE-1	HLA-A1	EADPTGHSY[b]
	HLA-Cw16	SAYGEPRKL
MAGE-3	HLA-A1	EVDPIGHLY[b]
	HLA-A24	IMPKAGLLI
	HLA-A2	FLWGPRALV
	HLA-B44	MEVDPIGHLY
BAGE	HLA-Cw16	AARAVFLAL
GAGE-1,2	HLA-Cw6	YRPRPRRY
NY-ESO-1	HLA-A31	ASGPGGGAPR
	HLA-A31	LAAQERRVPR[d]
	HLA-A2	SLLMWITQC
	HLA-A2	SLLMWITQCFL
	HLA-A2	QLSLLMWIT
	HLA-A31	ASGPGGGAPR

TABLE 3. *Continued*

Antigen	Ag-presenting MHC	Epitope
b. GnT-V	HLA-A2	VLPDVFIRC[c]
c. p15	HLA-A24	AYGLDFYIL
d. PRAME	HLA-A24	LYVDSLFFL
C. Tumor-specific unique peptides		
β-catenin	HLA-A24	SYLDSGIHF*
MUM-1	HLA-B44	EEKL*IVVLF[c]
CDK4	HLA-A2	AC*DPHSGHFV

Underline, nonoptimal amino acids in the HLA-A*0201 anchor positions;
*, mutation.
[a] Peptides recognized by many patients.
[b] Peptides used in clinical trials.
[c] Peptides from intron sequences.
[d] Peptides from alternative ORFs.

Melanosomal Proteins

Many cultured TIL recognize autologous melanoma cells in vitro as well as allogeneic melanoma cells and cultured melanocytes that share antigen-presenting HLA class I molecules, but do not recognize other tissue cells, suggesting that these TIL recognize nonmutated peptides derived from melanocyte-specific proteins [7]. Five melanosomal proteins (MART-1, gp100, tyrosinase, TRP-1, and TRP-2) were identified as antigens recognized by T cells. Tyrosinase, TRP-1, and TRP-2 are enzymes involved in the synthesis of melanins. Melanosomal proteins appear to represent highly immunogenic antigens. Langerhans cells, antigen-presenting dendritic cells in the skin, may efficiently take up the melanosomal proteins from melanoma cells similar to uptake of melanins by keratinocytes and efficiently induce T cells specific for these proteins.

MART-1 (melanoma antigen recognized by T cells-1) was isolated using an HLA-A2-restricted melanoma-reactive TIL [8]. MART-1 is an immunodominant melanoma antigen in HLA-A*0201 patients. It was recognized by the majority of HLA-A2-restricted melanoma-reactive TIL. A 9-mer peptide, AAGIGILTV (MART-1$_{27-35}$), represents an immunodominant T-cell epitope [9]. Although tumor regression was not observed in several patients who received the MART-1-reactive TIL, evidence of tumor regression was observed in some patients who were immunized with the MART-1 peptide in incomplete Freund's adjuvant (IFA) [10].

Gp100 was isolated using an HLA-A2-restricted TIL [11]. It was found to be also recognized by the murine monoclonal antibody HMB45, which was popularly used for the diagnosis of melanoma. Gp100 also appears to represent an immunodominant antigen in HLA-A*0201 patients. Three of ten identified gp100 epitopes, gp100$_{154-162}$ (KTWGQYWQV), gp100$_{209-217}$ (ITDQVPFSV), and gp100$_{280-288}$ (YLEPGPVTA), were recognized by TIL from many patients, suggesting that these were immunodominant common epitopes [12]. A significant correlation between gp100 recognition by TIL and clinical response to the adoptive transfer of these TIL was observed [12]. The results obtained from clinical trials using the gp100 antigen in the Surgery Branch (NCI) indicated that gp100 functioned as a tumor regression antigen. Because gp100 provides multiple epitopes for multiple HLAs including HLA-A1, -A2, -A3, and -A24, gp100 represents an attractive antigen for the development of effective immunotherapy [13,14].

Tyrosinase was isolated using HLA-A*2402- and HLA-A*0201-restricted melanoma-reactive CTL [15]. Tyrosinase was also recognized by HLA-DR4-restricted CD4+ melanoma-reactive TIL [16]. Adoptive transfer of one of the HLA-A24-restricted tyrosinase-reactive TIL into an autologous patient resulted in complete regression of tumor, and immunization with the tyrosinase peptide induced tumor regression. TRP1 was isolated using an HLA-A31-restricted melanoma-reactive TIL [17]. It was previously shown to be recognized by IgG antibodies in the serum of a melanoma patient. Transfer of the TRP1-reactive TIL to the autologous patient resulted in partial tumor regression. TRP2 was identified to be recognized by one of the HLA-A31-restricted TIL clones [18]. The HLA-A31-binding T-cell epitope, LLGPGRPYR, was found to bind to HLA-A33 and to be recognized by HLA-A33 restricted TIL [19]. An HLA-A*0201-binding epitope, SVYDFFVWL, was identified by screening 21 HLA-A2-binding TRP2 peptides using the in vitro CTL induction method [20]. This peptide was found to be the same as the epitope recognized by H-2K^b restricted murine B16 melanoma-reactive CTL [21]. This TRP2/B16 murine tumor model may be useful for evaluation of immunotherapy approaches using melanosomal proteins.

All epitopes identified in the melanosomal antigens are nonmutated self-peptides. The HLA-binding affinity of some of these epitopes was found to be relatively low [12]. Because MHC-binding affinity influences the peptide density on the cell surface, these epitopes appear to be subdominant or cryptic epitopes that are expressed at very low density on the surface of melanocytes [22]. Absence of optimal anchor amino acids at the primary anchor positions appears to be responsible for their low MHC-binding affinity (Table 3) [12]. Immunological tolerance may not be completely induced in the CTL specific for these epitopes. These CTL are silent in a normal individual; however, they may be activated in patients with melanoma. Tissue destruction and inflammation at the tumor site may induce T-cell activation directly or indirectly through the production of various cytokines and the increased expression of antigen, MHC, or accessory molecules on the tumor cell surface. It was reported that melanosomal proteins were frequently misfolded in melanoma cells and were subsequently ubiquitinated and processed in proteasomes. Thus, melanosomal peptides may be presented by MHC class I molecules in a higher amount on the surface of melanoma cells. These in vivo primed T cells may be further expanded as tumor-reactive T cells in in vitro culture with IL-2.

Melanosomal proteins may be useful in the immunotherapy for patients with melanoma [23]. However, tumor cells may escape from T-cell recognition through a loss of these antigens, because melanosomal proteins are not important for tumor cells to proliferate and survive. In fact, 5%–20% of patients with advanced metastases were found to have metastases that lost expression of MART-1 or gp100 [24]. The induction of immune responses against melanosomal proteins may induce destruction of melanocytes. Vitiligo was observed in some patients who received IL-2-based immunotherapies in the Surgery Branch, NCI [25]. However, no ophthalmic problem caused by destruction of melanocytes in uvea or pigmented epithelial cells in retina has been observed. Because susceptibility of normal and tumor cells to immune responses may differ because of different tissue structure, inflammatory status, and epitope density, immune responses against self-proteins may not always develope autoimmune adverse effects.

Cancer-Testis Antigens

Proteins expressed on various cancers and normal testis were found to be melanoma antigens recognized by T cells. MAGE-1 was the first identified as a cancer-testis antigen [26]. We have isolated NY-ESO-I as an antigen for HLA-A31-restricted, melanoma-reactive TIL [27]. Cancer-testis antigens appear to be tumor-specific, shared antigens as they are expressed on spermatogonia and spermatocytes that do not express MHC class I molecules. Tumor regression was observed in some patients after immunization with the HLA-A1-binding MAGE-3 peptide. However, the role of this antigen in in vivo tumor regression has not been clarified, because no MAGE-3-specific CTL was induced from the immunized patients [28].

Peptides Derived from Mutation of Widely Expressed Proteins

A mutated peptide of β-catenin was identified as an antigen recognized by HLA-A24-restricted TIL [29]. A serine to phenylalanine transition by C-to-T mutation generated an epitope peptide capable of binding to HLA-A24. Administration of this CTL into the autologous patient was associated with complete tumor regression. Eight of 26 melanoma cell lines were found to overexpress the β-catenin protein because of either mutations or unusual splicing of β-catenin or inactivation of APC [30]. Thus, defective β-catenin regulation appears to be involved in development of melanoma. Similar findings were observed in the mutated CDK4 melanoma antigen [31].

Mutated antigens are truly tumor specific and may be ideal targets for immunotherapy. These were isolated using T cells from patients who had a good prognosis after treatment, suggesting that immune responses to these mutated peptides might be involved in tumor regression. Antigen loss variants may not easily develop, because these mutated molecules may be important for tumor growth. However, it is not clear whether immunogenic mutated epitopes are presented by MHC molecules on the surface of most tumor cells. Tumor cells that expressed highly immunogenic peptides may have already been eliminated before being clinically observed. It is also difficult to apply unique epitopes for immunotherapy unless mutations are common or more efficient techniques can be developed for the isolation of tumor antigens.

The Mechanisms for Generation of T-Cell Epitopes on Melanoma Cells

Isolation of genes encoding melanoma antigens revealed a variety of mechanisms to generate T-cell epitopes on growing melanoma cells (Table 4). Many melanoma epitopes were nonmutated peptides derived from regular open reading frames (ORFs). Nonmutated epitopes from melanosomal proteins appear to be subdominant/cryptic epitopes. A mutated peptide derived from

TABLE 4. Mechanisms for generation of T-cell epitopes on melanoma cells

Mechanism	Antigen
A. Peptides from normal ORFs	MART-1/melanA*, gp100*, tyrosinase*,TRP2*, p15*, MAGE1, MAGE3, BAGE, GAGE, NY-ESO-I*
B. Peptides from alternative ORFs	TRP1*, NY-ESO-I*
C. Peptides from genes with mutation	β-catenin*, CDK4, MUM-1
D. Peptides from intron sequence in incompletely spliced RNAs	gp100*, MUM-1
E. Peptides modified after translation	Tyrosinase
F. Peptides from transcription by cryptic promoter	N-Acetyl glucosaminyl transferase-V

* Antigens isolated with TIL (tumor-infiltrating lymphocytes) in the Surgery Branch, NCI.

widely expressed proteins including β-catenin and CDK4 was found to be a tumor-specific antigen. The TRP1 epitope was derived from a short polypeptide encoded by the third ORF different from that encoding the DHICA oxidase [32]. One of the NY-ESO-I epitopes was also derived from an alternative ORF [27]. The HLA-A24-binding gp100 epitope was derived from an intron sequence in an incompletely spliced mRNA that appeared to be present in low amount in melanoma as well as in cultured melanocytes [33]. One tyrosinase epitope was suggested to have posttranslational modification because none of the synthetic peptides in the region containing the epitope was recognized by T cells.

Development of Antigen-Specific Immunotherapy for Melanoma

A variety of antigen-specific immunotherapies without identification of tumor antigens has recently been proposed, and some clinical trials have already taken place (Table 5). These protocols may be able to immunize with multiple unknown tumor antigens, including unique mutated antigens.

TABLE 5. Antigen-specific immunotherapies for melanoma

I. Immunotherapies without identification of tumor antigens
A. Passive immunotherapy
  1. Administration of melanoma-reactive T cells generated in vitro by stimulation with melanoma cells
B. Active immunization
  1. Melanoma cells transfected with genes encoding cytokines, costimulatory molecules, or foreign antigens
  2. Dendritic cells pulsed with peptides, proteins, or RNA extracted from melanoma cells
  3. Dendritic cells fused with melanoma cells
  4. Heat shock proteins extracted from melanoma cells
II. Immunotherapies with the identified tumor antigens
A. Passive immunotherapy
  1. Administration of melanoma-reactive T cells generated in vitro by stimulation with peptides
B. Active immunization
  1. Peptides or whole proteins (with adjuvants and cytokines, combined with lipids or liposomes, with heat shock proteins)
  2. Recombinant viruses containing antigen genes (adenovirus, fowlpox virus, vaccinia virus, retrovirus)
  3. Plasmid DNAs containing antigen genes (intramuscular injection or with gene gun)
  4. Recombinant bacteria containing antigen genes (BCG, *Salmonella*, *Listeria*)
  5. Dendritic cells pulsed with antigenic peptides or proteins, or transfected with antigen genes

However, it is unclear whether a low amount of antigens obtained from tumor samples can mount T-cell responses sufficient to cause tumor regression. In contrast, a large amount of the identified antigens is available for immunotherapy, although only a limited number of antigens has been so far identified and a therapeutic role of any identified antigens has not been established. It is also difficult to use autologous tumor-specific antigens. However, these antigens are readily available for testing their tumor rejection ability in clinical trials. In particular, antigens presented by multiple HLA molecules or by the frequently expressed HLA including HLA-A2, which is expressed in about 50% of Caucasians, can be applied in many patients.

Efficient immunization protocols such as methods to allow high-density expression of tumor epitopes on the cell surface of professional antigen-presenting cells (APC), need to be developed (Table 6). Peptides may be used for immunization in conjunction with adjuvants and cytokines. Cytokines such as IL-2, IL-12, GM-CSF, or interferon may enhance the efficacy of immunotherapies by improving immunization as well as by enhancing tumor recognition at the effector phase through the increase of epitope–MHC complexes on the tumor cell surface. It needs to be evaluated which adjuvants and cytokines are most effective in the immunization with tumor antigens.

Tumor regression was observed in some patients in the phase I clinical trials using the peptides from MART-1, gp100, or tyrosinase, with either an

TABLE 6.   Phase I immunotherapy trials using the identified melanoma antigens in Surgery Branch, NCI

MART-1$_{27}$ peptide in IFA (alone or with IL12)

gp100 peptides$_{(154,209,280)}$ in IFA

Modified gp100$_{209(210M)}$ peptide in IFA (alone or with IL2, IL12, GM-CSF, or Flt3L)

Modified gp100$_{280(288V)}$ peptide in IFA

Four peptides (MART-1$_{27}$, tyrosinase$_{368(370D)}$, gp100$_{209(210M)}$, gp100 peptide$_{280(288V)}$) in IFA (alone or with IL2)

Recombinant adenovirus encoding MART-1 (alone or with IL-2)

Recombinant adenovirus encoding gp100 (alone or with IL-2)

Recombinant fowlpox virus encoding MART-1 (alone or with IL-2)

Recombinant fowlpox virus encoding gp100 (alone or with IL-2)

Recombinant vaccinia virus encoding gp100 (alone or with IL-2)

Recombinant vaccinia virus encoding MART-1 (alone or with IL-2)

Autologous dendritic cells pulsed with the MART-1 and gp100$_{209(210M)}$ peptides

Adoptive transfer of PBL sensitized in vitro to the gp100$_{209(210M)}$ peptide

Adoptive transfer of CTL clones with high-avidity TCR, isolated from PBL sensitized in vitro to the gp100$_{209(210M)}$ peptide

---

IFA, incomplete Freund's adjuvant; PBL, peripheral blood lymphocytis; CTL, cytotoxic T lymphocyte; TCR, T-cell receptor.

IFA or GM-CSF [10,34]. The observation that melanoma metastases that lost expression of melanosomal antigens grew progressively, while other multiple metastases regressed in some patients, suggests that immune responses against the immunized antigens might cause tumor regression. Immune augmentation was detected in most patients immunized with the MART-1 or gp100 peptides by comparing in vitro CTL induction before and after the immunizations [10,34]. However, CTL precursor analysis demonstrated the precursor frequency was less than 1 in 30000 in peripheral blood even after the immunization. Because many melanoma epitopes have a relatively low MHC-binding affinity and MHC-binding affinity correlates with immunogenicity of peptides, we attempted to generate more immunogenic peptides by increasing MHC-binding affinity. By replacing amino acids at the anchor residue, gp100 peptides with tenfold higher HLA-A2-binding affinity could be generated. The high-binding $gp100_{209(210M)}$ peptide (threonine was replaced by methionine at P2) could induce melanoma reactive CTL more efficiently in vitro and in vivo than the native epitope [35]. The CTL precursor frequency was increased to about 1 in 3000. A clinical trial using $gp100_{209(210M)}$ along with IFA and IL-2 was performed in the Surgery Branch (NCI) and resulted in either complete regression (CR) or partial regression (PR) in 13 of 31 patients (42% response rate) [36].

Because IL-2 alone had antimelanoma effects (15% in another trial), immunization with the peptide might be responsible for the increased response rate. This antitumor effect needs to be confirmed in the phase II clinical trial, and the role of the gp100 peptide in tumor regression needs to be evaluated by analysis of T cells infiltrating in the tumors. Augmentation of systemic immunity against antigens may not be sufficient for tumor rejection. IL-2 might help antitumor T cells functioning in the tumors through capillary leak syndrome, which might lead antitumor T cells infiltrating into tumors, as well as through production of various cytokines. A clinical trial of the immunization with multiple peptides (MART-1$_{27}$, tyrosinase$_{368(370D)}$, $gp100_{209(210M)}$, $gp100_{280(288V)}$) is in progress in the Surgery Branch.

Administration of dendritic cells (DC) pulsed with peptides was demonstrated to immunize more efficiently than direct peptide injection in murine tumor models. Injection of DC pulsed with tumor lysates or synthetic peptides into lymph nodes along with KLH was reported to result in a 31% response rate in patients with melanoma [37]. However, intravenous administration of DC pulsed with the gp100 and MART-1 peptides did not demonstrate effective immunization or antitumor effects. Immunization protocols with DC needs to be optimized by considering mode of administration and maturation status of DC.

Immunization with recombinant viruses containing tumor antigen genes is an efficient method in murine tumor models. In addition to the high-level expression of antigens, recombinant viruses may provide multiple epitopes for both CD4+ and CD8+ T cells. However, in the immunization trials with recombinant vaccinia virus, fowlpox virus and adenovirus containing MART-1 or gp100, effective immunization or antitumor effect has not been observed [38]. High-titer neutralizing antibodies against viruses were detected in patients, indicating that antiviral immune responses inhibited immunization of tumor antigens. Although immunization with the plasmids containing antigen genes may be less efficient for induction of T-cell responses, it may be useful to boost immune responses by repeated adminis-tration. A clinical protocol using the gp100 plasmid has begun in the Surgery Branch (NCI).

Because effective immunization may not be accomplished in immunosup-pressed patients, passive transfer of antitumor effector T cells that are generated under less suppressive in vitro conditions may provide better antitumor activity. Melanoma-reactive CTL with higher antitumor activity than the conventional TIL can be generated from PBL of patients by in vitro stimulation with the melanoma peptides, particularly from the patients preimmunized with the tumor antigens [39].

Tumor cells may escape from T-cell responses through a variety of mech-anisms (Table 7). In 155 metastases from advanced melanoma patients in the Surgery Branch (NCI), 8%, 21%, or 6% was found to have lost expression of MART-1, gp100, or HLA-A2, respectively, using an immunohistochemical technique [24]. It is important to understand the mechanisms of the tumor escape ae well as the actual frequency of the occurrence in patients to improve immunotherapy in the future. Multiple antigens may be used to prevent tumor escape through the emergence of antigen loss variants. Other types of thera-pies including chemotherapy are probably necessary for the eradication of MHC loss variants. Additional procedures may also be necessary to correct negative immune influence against antitumor immunity.

TABLE 7. Mechanisms for tumor escape from T-cell responses

1. Loss of recognition molecules necessary for T-cell recognition
   Antigen
   MHC heavy chain, β_2-microglobulin
   TAP, LMP
2. Negative immune regulation against antitumor immune responses
   Tolerance induction (deletion, anergy)
   Suppressive factors from tumor cells or host cells
   Fas ligand expression on tumor cells
   Altered T-cell signaling
   Th1/Th2 shift

Concluding Remarks

Preliminary results from the phase I clinical trials suggest that some of the identified melanoma antigens function as tumor rejection antigens, although the role of T-cell responses in in vivo tumor rejection has not yet been established in most antigens. More efficient immunization methods and maneuvers to overcome local blockade in tumors need to be developed on the basis of the mechanisims of tumor escape. Clinically, it is very important to identify parameters to predict which patients will respond to immunotherapy. Findings obtained from these research on melanoma may help development of immunotherapy for other cancers.

Acknowledgments. A part of this study was supported by grants-in-aid from the Ministry of Education, Science, Sport and Culture of Japan (10470264, 10557083, 10670175, 10671492, 9255106), grant-in-aid for Cancer Research (11–7) from the Ministry of Health and Welfare, Ciba-Geigy Foundation (Japan) for the Promotion of Science, The Vehicle Racing Commemorative Foundation, The Naito Foundation, Japanese Foundation for Multidisciplinary Treatment of Cancer, Terumo Life Science Foundation, The Keio University Medical Science Fund, and Keio University Special Grant-in-Aid for Innovative Collaborative Research Projects. This study was also performed in the Surgery Branch, National Cancer Institute, National Institutes of Health, Bethesda, MD, USA. I thank all the members in the Surgery Branch, especially Steven A. Rosenberg, for continuous support, and Paul F. Robbins and RongFu Wang for the project on the isolation and characterization of melanoma antigens.

References

1. Rosenberg S, Yang J, White D, Steinberg S (1998) Durability of complete responses in patients with metastatic cancer treated with high-dose interleukin-2. Ann Surg 228:307–319
2. Rosenberg SA, Yannelli JR, Yang JC, Topalian SL, Schwartzentruber DJ, Weber JS, Parkinson DR, Seipp CA, Einhorn JH, White DE (1995) Treatment of patients with metastatic melanojma with autologous tumor-infiltrating lymphocytes and interleukin 2. J Natl Cancer Inst 86:1159–1166
3. Kawakami Y, Robbins PF, Wang RF, Rosenberg SA (1996) Identification of melanoma antigens recognized by T lymphocytes and their use in the immunotherapy of cancer. In: Devita VT, Hellman S, Rosenberg SA (eds) Principle and practice of oncology: update. Lippincott-Raven, Philadelphia, pp 1–20
4. Kawakami Y (1998) Immunotherapy using T-cell-defined tumor antigens for melanoma. Microbiol Immunol 42:803–813
5. Old LJ, Chen YT (1998) New paths in human cancer serology. J Exp Med 187:1163–1167

6. Sahin U, Tureci O, Pfreundschuh M (1997) Serological identification of human tumor antigens. Curr Opin Immunol 9:709–716

7. Kawakami Y, Zakut R, Topalian SL, Stotter H, Rosenberg SA (1992) Shared human melanoma antigens. Recognition by tumor infiltrating lymphocytes in HLA-A2.1 transfected melanomas. J Immunol 148:638–643

8. Kawakami Y, Eliyahu S, Delgaldo CH, Robbins PF, Rivoltini L, Topalian SL, Miki T, Rosenberg SA (1994) Cloning of the gene coding for a shared human melanoma antigen recognized by autologous T cells infiltrating into tumor. Proc Natl Acad Sci USA 91:3515–3519

9. Kawakami Y, Eliyahu S, Sakaguchi K, Robbins PF, Rivoltini L, Yannelli JB, Appella E, Rosenberg SA (1994) Identification of the immunodominant peptides of the MART-1 human melanoma antigen recognized by the majority of HLA-A2 restricted tumor infiltrating lymphocytes. J Exp Med 180:347–352

10. Cormier JN, Salgaller ML, Prevette T, Barracchini KC, Rivoltini L, Restifo NP, Rpsenberg SA, Marincola FM (1997) Enhancement of cellular immunity in melanoma patients immunized with a peptide from MART-1/Melan A. Cancer J Sci Am 3:37–44

11. Kawakami Y, Eliyahu S, Delgado CH, Robbins PF, Sakaguchi K, Appella E, Yannelli JR, Adema GJ, Miki T, Rosenberg SA (1994) Identification of a human melanoma antigen recognized by tumor infiltrating lymphocytes associated with in vivo tumor rejection. Proc Natl Acad Sci USA 91:6458–6462

12. Kawakami Y, Eliyahu S, Jennings C, Sakaguchi K, Kang X-Q, Southwood S, Robbins PF, Sette A, Appella E, Rosenberg SA (1995) Recognition of multiple epitopes in the human melanoma antigen gp100 associated with in vivo tumor regression. J Immunol 154:3961–3968

13. Kawakami Y, Robbins PF, Wang X, Tupesis JT, Fitzgerald E, Li YF, El-Gamil M, Matthews WJ, Parkhurst PR, Kang X, Sakaguchi K, Appella E, Rosenberg SA (1998) Identification of new melanoma epitopes on melanosomal proteins recognized by tumor infiltrating T lymphocytes restricted by HLA-A1, -A2, and -A3 alleles. J Immunol 161:6985–6992

14. Kawakami Y, Dang N, Wang X, Tupesis JT, Robbins PF, Wunderlich JR, Yannelli JR, Rosenberg SA (1999) Recognition of shared melanoma antigens in association with major HLA-A alleles by tumor infiltrating T lymphocytes from 123 patients with melanoma. J Immunother (in press)

15. Robbins PF, El-Gamil M, Kawakami Y, Stevens E, Yannelli J, Rosenberg SA (1994) Recognition of tyrosinase by tumor infiltrating lymphocytes from a patient responding to immunotherapy. Cancer Res 54:3124–3126

16. Topalian SL, Gonzales MI, Parkhurst M, Li YF, Southwood S, Sette A, Rosenberg SA, Robbins PF (1996) Melanoma-specific CD4+ T cells recognize nonmutated HLA-DR-restricted tyrosinase epitopes. J Exp Med 183:1965–1971

17. Wang RF, Robbins PF, Kawakami Y, Kang XQ, Rosenberg SA (1995) Identification of a gene encoding a melanoma tumor antigen recognized by HLA-A31-restricted tumor-infiltrating lymphocytes. J Exp Med 181:799–804

18. Wang RF, Appella E, Kawakami Y, Kang X, Rosenberg SA (1996) Identification of TRP2 as a human tumor antigen recognized by cytotoxic T lymphocytes. J Exp Med 184:2207–2216

19. Wang RF, Johnson SL, Southwood S, Sette A, Rosenberg SA (1998) Recognition of an antigenic peptide derived from tyrosinase-related protein-2 by CTL in the context of HLA-A31 and -A33. J Immunol 160:890–897

20. Parkhurst PR, Fitzgerald E, Southwood S, Sette A, Rosenberg SA, Kawakami Y (1998) Identification of a shared HLA-A*0201 restricted T cell epitope from the melanoma antigen tyrosinase related protein 2 (TRP2). Cancer Res 58:4895–4901

21. Bloom MB, Lalley DP, Robbins PF, Li Y, El-Gamil M, Rosenberg SA, Yang JC (1997) Identification of TRP2 as a tumor rejection antigen for the B16 melanoma. J Exp Med 185:453–459

22. Kawakami Y, Rosenberg S (1995) T-cell recognition of self peptides as tumor rejection antigens. Immunol Res 15:179–190

23. Kawakami Y, Robbins P, Wang R, Parkhurst M, Kang X, Rosenberg S (1998) The use of melanosomal proteins in the immunotherapy of melanoma. J Immnother 21:237–246

24. Cormier J, Hijazi Y, Abati A, Fetsch P, Bettinotti M, Steinberg S, Rosenberg S, Marincola M (1998) Heterogeneous expression of melanoma-associated antigens and HLA-A2 in metastatic melanoma in vivo. Int J Cancer 75:517–524

25. Rosenberg SA, White DE (1996) Vitiligo in patients with melanoma: normal tissue antigens can be target for cancer immunotherapy. J Immunother 19:81–84

26. Van der Bruggen P, Traversari C, Chomez P, Lurquin C, DePlaen E, Van Den Eynde B, Knuth A, Boon T (1991) A gene encoding an antigen recognized by cytolytic T lymphocytes on a human melanoma. Science 254:1643–1647

27. Wang R-F, Johnston S, Zeng G, Topalian S, Schwartzentruber D, Rosenberg S (1998) A breast and melanoma-shared tumor antigen: T-cell responses to antigenic peptides translated from different open reading frames. J Immunol 161:3596–3606

28. Marchand M, Baren N, Weynants P, Brichard D, Dreno B, Tessier MH, Rankin E, Parmiani G, Arienti F, Humblet Y, Bourlond A, Vanwijck R, Lienard D, Beauduin M, Dietrich PY, Russo V, Kerger J, Masucci G, Jager E, De Greve J, Atzpodien J, Brasseur F, Coulie PG, Van der Bruggen P, Boon T (1999) Tumor regressions observed in patients with metastatic melanoma treated with an antigenic peptide encoded by MAGE-3 and presented by HLA-A1. Int J Cancer 80:219–230

29. Robbins PF, El-Gamil M, Li YF, Kawakami Y, Loftus D, Appella E, Rosenberg SA (1996) A mutated β-catenin gene encodes a melanoma-specific antigen recognized by tumor infiltrating lymphocytes. J Exp Med 183:1185–1192

30. Rubinfeld B, Robbins P, El-Gamil M, Albert I, Prfiri E, Polakis P (1997) Stabilization of beta-catenin by genetic defects in melanoma cell lines. Science 275:1790–1792

31. Wolfel T, Hauer M, Schneider J, Serrano M, Wolfel C, Klehmann-Hieb E, De Plaen E, Hankeln T, Meyer Zum Buschenfelde K-H, Beach D (1995) A p16INK4a-insensitive CDK4 mutant targeted by cytolytic T lymphocytes in a human melanoma. Science 269:1281–1284

32. Wang R-F, Parkhurst M, Kawakami Y, Robbins PF, Rosenberg SA (1996) Utilization of an alternative open reading frame of a normal gene in generating a novel human cancer antigen. J Exp Med 183:1131–1140

33. Robbins PF, El-Gamil M, Li YF, Fitzgerald E, Kawakami Y, Rosenberg SA (1997) The intronic region of an incompletely-spliced gp100 gene transcript encodes an epitope recognized by melanoma reactive tumor infiltrating lymphocytes. J Immunol 159:303–308

34. Jager E, Ringhoffer M, Karbach J, Arand M, Oesch F, Knuth A (1996) Inverse relationship of melanocyte differentiation antigen expression in melanoma tissues and CD8+ cytotoxic-T-cell responses: evidence for immunoselection of antigen-loss variants in vivo. Int J Cancer 66:470–476

35. Parkhurst MR, Salgaller M, Southwood S, Robbins P, Sette A, Rosenberg SA, Kawakami Y (1996) Improved induction of melanoma reactive CTL with peptides from the

melanoma antigen gp100 modified at HLA-A0201 binding residues. J Immunol 157:2539–2548

36. Rosenberg S, Yang J, Schwartzentruber D, Hwu P, Marincola F, Topalian S, Restifo N, Dudley M, Schwarz S, Spiess P, Wunderlich J, Parkhurst M, Kawakami Y, Seipp C, Einhorn J, White D (1998) Immunologic and therapeutic evaluation of a synthetic peptide vaccine for the treatment of patients with metastatic melanoma. Nat Med 4:321–327

37. Nestle F, Alijagic S, Gilliet M, Sun Y, Grabbe S, Dummer R, Burg G, Schadendorf D (1998) Vaccination of melanoma patients with peptide- or tumor lysate-pulsed dendritic cells. Nat Med 4:328–332

38. Rosenberg S, Zhai Y, Yang J, Schwartzentruber D, Hwu P, Mrincola F, Topalian S, Restifo N, Seipp C, Einhorn J, Roberts B, White D (1998) Immunizing patients with metastatic melanoma using recombinant adenoviruses encoding MART-1 or gp100 melanoma antigens. J Natl Cancer Inst 90:1894–1900

39. Salgaller M, Marincola F, Rivoltini L, Kawakami Y, Rosenberg S (1995) Recognition of multiple epitopes in the human melanoma antigen gp100 by antigen specific peripheral blood lymphocytes stimulated with synthetic peptides. Cancer Res 55:4972–4979

Part 2
Hematopoietic Stem Cells
Biology and Clinical Application

Part 2

Isolation and Characterization of CD34-Low/Negative Mouse Hematopoietic Stem Cells

HIROMITSU NAKAUCHI, MASATAKE OSAWA*, KAZUHIRO SUDO, and HIDEO EMA

Summary. We have previously reported that, in adult mouse bone marrow, CD34$^{low/-}$ c-Kit$^+$ Sca-1$^+$ lineage markers negative (Lin$^-$) (CD34$^-$KSL) cells represent hematopoietic stem cells with long-term marrow repopulating ability whereas CD34$^+$ c-Kit$^+$ Sca-1$^+$ Lin$^-$ (CD34$^+$KSL) cells are progenitors with short-term reconstitution capacity. To characterize these two populations of cells further, relative expression of various genes was examined by RT-PCR. In CD34$^-$KSL cells, most cytokine receptor genes were not expressed with the exception of IL2Rγ and AIC-2B. In contrast, expression of all cytokine receptor genes examined except IL-2Rα, IL-7Rα, and IL9Rα chains were found in CD34$^+$KSL cells. Cell cycle studies revealed only 3% of CD34$^-$KSL cells and 26% of CD34$^+$KSL cells are dividing at a given time. Long-term BrdU administration study demonstrated, however, that majority of CD34$^-$KSL cells contribute to hemopoiesis by dividing very slowly. Furthermore, analysis of aged mice revealed more than tenfold increase in absolute number of CD34$^-$KSL cells. Those CD34$^-$KSL cells in aged mice appeared to include HPP-CFC at an equivalent frequency with those in younger mice. These data support our previous notion that CD34$^-$KSL cells are at higher rank in hematopoietic hierarchy than CD34$^+$KSL cells. In addition, our results provide important clues for cell therapy and gene therapy targeting hematopoietic stem cells.

Key words. Hematopoietic stem cells, CD34, Cell cycle, Aging, Bone marrow transplantation

Institute of Basic Medical Sciences and Center for TARA, University of Tsukuba, and CREST (JST), 1-1-1 Ten-nodai, Tsukuba, Ibaraki 305-8575, Japan
* *Present address*: KIRIN Pharmaceutical Research Laboratory, Gunma 370-1295, Japan

Introduction

Many adult tissues contain a population of cells that are capable of supplying all types of cells in that particular tissue. Because of multilineage differentiation and their self-renewing potential, those cells are called stem cells, and have been considered as a potential target for cell therapy and gene therapy. It is therefore critically important to identify them and to understand the mechanism of their differentiation and self-renewal. During the past 10 years, we have been working on isolation and characterization of hematopoietic stem cells (HSCs). HSCs in the bone marrow supply all blood cells throughout life by making use of their self-renewal and multilineage differentiation capabilities. Thus, HSCs are the essence of bone marrow transplantation and regeneration of blood cells. Despite the crucial role of HSCs in normal hematopoiesis as well as in clinical bone marrow (BM) transplantation, our knowledge of their physical characteristics and the mechanisms that control their proliferation and differentiation remain elusive.

Although isolation of HSCs is essential for further quantitative and molecular biological analyses of differentiation and self-renewal capabilities, it has been difficult because of their paucity in the BM. However, recent progress in cell separation techniques and the development of monoclonal antibodies have enabled the isolation of HSC from mammalian bone marrow cells [1]. For example, a monoclonal antibody to c-Kit clearly showed that this receptor tyrosine kinase is expressed on hematopoietic progenitor cells and can be used as a marker for HSC enrichment [2,3]. Three-color FACS analysis and cell-sorting experiments revealed that long-term marrow-repopulating ability (LMRA) is exclusively enriched in the c-Kit⁻Sca-1⁻fraction among mouse bone marrow lineage marker-negative (Lin⁻) cells [4]. However, a minimum of 30 c-Kit⁺Sca-1⁺ Lin⁻ cells were required to radioprotect lethally irradiated mice. We therefore needed another monoclonal antibody that could subdivide c-Kit⁺Sca-1⁺ Lin⁻ cells for further enrichment.

In humans, CD34 has been used as a marker for HSCs because all colony-forming activity is found in the CD34-positive fraction. Clinical transplantation as well as primate studies that used enriched CD34⁺ BM cells also indicated the presence of HSCs with long-term marrow-repopulating ability within this fraction [5]. CD34 is a cell-surface sialomucin-like adhesion molecule that is expressed on 1%–3% of BM cells. After isolation of the human CD34 gene [6,7], the mouse homologue (mCD34) was isolated by cross-hybridization [8,9]. With the expectation that mCD34 is expressed in mouse HSCs, a rat monoclonal antibody was raised against recombinant mCD34. This antibody (clone 49E8) reacted with 2.5% ± 0.5% of adult mouse BM cells that were mostly c-Kit⁺ and Lin⁻, suggesting its expression in primitive hematopoietic cells. Using this mCD34 monoclonal antibody, we were indeed able to further subdivide c-Kit$^+$Sca-1$^+$ Lin$^-$ (KSL) cells. Day-12 CFU-S and CFU-C were found mainly in the CD34$^+$KSL fraction. However, contrary to our expectation, those cells in the CD34-positive fraction did not show LMRA. In adult mouse BM, HSCs with LMRA have a phenotype of CD34$^{low/-}$ c-Kit$^+$ Sca-1$^+$ Lin$^-$ (CD34$^-$KSL), whereas the CD34$^+$KSL fraction contains hematopoietic progenitors with short-term reconstitution ability [10]. In this chapter our recent results of in vivo as well as in vitro characterization of those highly enriched hematopoietic stem/progenitor populations are described.

Both CD34$^-$KSL and CD34$^+$KSL Subpopulations Are Required for Successful Bone Marrow Transplantation

To study their radioprotective ability, KSL cells were fractionated into mCD34-negative, CD34^{low}, and CD34-positive subpopulations according to their mCD34 expression by FACS. As shown in Fig. 1, although injection of 100 KSL cells was shown to radioprotect a lethally irradiated mouse [3], injection of 300 cells from either the CD34$^-$ or CD34$^+$KSL subpopulation alone showed poor radioprotection. When cells of both fractions were cotransplanted, rescue of lethally irradiated mice was observed. These results indicate that when a limited number of HSCs are used for transplantation, both commit-

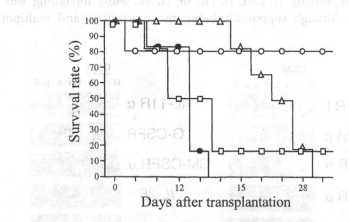

FIG. 1. Radioprotective ability of c-Kit$^+$ Sca-1$^+$ Lin$^-$ bone marrow subpopulations. Bone marrow KSL cells were subdivided into three fractions according to the expression of mCD34 (Fr1; CD34$^-$, Fr2; CD34low, and Fr3; CD34$^+$). Survival rate of recipient mice after transfer of cells is indicated. Groups of 30 mice were lethally irradiated (9.5 Gy) and injected intravenously with either 300 CD34$^-$KSL cells (Fr. 1; *open squares*), CD34$^+$KSL cells (Fr. 3; *open triangles*), or CD34$^-$KSL cells plus CD34$^+$KSL cells (*open circles*) in four separate experiments. *Closed circles* indicate negative control (without injection)

ted progenitor cells that provide initial engraftment and HSCs, which are responsible for delayed but durable engraftment, are necessary for successful long-term reconstitution.

Expression of Various Genes in Purified Mouse HSCs

To examine the expression of various genes in highly enriched hematopoietic stem cell subpopulations, mRNAs were obtained from sorted cells and RT-PCR analysis was performed. Because these cells are rare in bone marrow, we sorted 2000 cells from each subpopulation. After first-strand synthesis, polyA tailing reaction was performed by terminal deoxynucleotidyl transferase (TdT). PCR amplification was then performed with 1 mM (A/C/G)-dT18-R1-R0 primer as described by Brady et al. [11].

Using these amplified cDNA, PCR reactions were performed with primers specific for various genes. Preliminary studies using primers specific for housekeeping genes confirmed uniform amplification of cDNA obtained from CD34$^-$, CD34$^+$, and Lin$^-$ cells. Figure 2 demonstrates the relative expression of various cytokine receptor genes in the three cell populations. Lin$^-$ cells do not contain mature blood cells but include a number of lineage-committed progenitor cells. Thus, all cytokine receptors tested were found to be expressed in Lin$^-$ cells. CD34$^+$ progenitor cells express all except IL-2Rα, IL-7R, and IL-9R. We can assume that CD34$^+$ progenitor fraction cells are more primitive than the Lin$^-$ cell population that contains committed progenitor cells expressing IL-2Rα, IL-7R, or IL-9R. Most intriguing was the finding that, although supposedly being most primitive and multipotent,

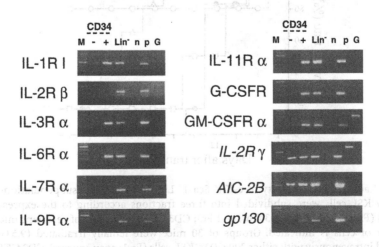

FIG. 2. Expression of cytokine receptor genes in CD34$^-$KSL, CD34$^+$KSL, Lin$^-$ cells. M, size marker; −, negative control; +, positive control; G, genomic DNA

CD34-negative HSCs did not express most cytokine receptors including IL-3R, IL-6R, and GM-CSFR. These data are in accord with our previous observation that CD34⁻KSL cells do not form colonies in the presence of IL-3 alone, but CD34⁺KSL cells do [10].

The fact that CD34-negative HSCs do not respond to single cytokines and require the presence of SCF and one other cytokine to form colonies may indicate that SCF mediates signals somehow act to induce expression of other cytokine receptors. By demonstrating expression of multiple cytokine receptors in the hematopoietic progenitor cells, Hu et al. proposed a promiscuous model in which lineage commitment is prefaced by a phase of multilineage locus activation [12]. From our results, however, this multilineage locus activation takes place not at HSC level, but most probably at the CD34⁺ progenitor cell stage.

Although most cytokine receptor α-chains were not expressed, mRNA for common signal transducing molecules such as IL-2Rγ chain and AIC-2B were detectable. As far as we examined, all known IL-2R γ-chain-associated α- or β-chains were not expressed. There may be as yet unknown α-chains expressed in HSCs. It has been reported, however, that IL-2R γ-chain knockout mice showed minimum hematopoietic abnormalities [13].

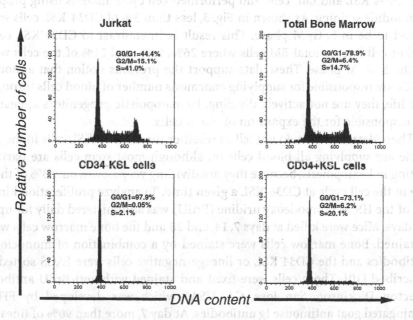

FIG. 3. Cell-cycle kinetics of Jurkat T cell line (*upper left*), total mouse bone marrow cells (*upper right*), CD34⁻KSL (*lower left*), and CD34⁺KSL (*lower right*) cells. Cells were fixed by paraformaldehyde, stained by propidium iodide, and subjected to FACS analysis. Relative frequency of cells in G_0/G_1, G_2/M, and S phase

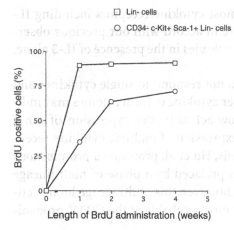

□ Lin- cells
○ CD34- c-Kit+ Sca-1+ Lin- cells

Length of BrdU administration (weeks)

FIG. 4. Long-term BrdU incorporation study. Proliferation kinetics of CD34⁻KSL and Lin– cells was analyzed using the long-term BrdU incorporation method. Mice were administered with BrdU daily for up to 28 days. Mice were killed at days 7, 14, and 28, and Lin– (*squares*) as well as CD34⁻KSL (*circles*) cells were obtained using FACS. Incorporation of BrdU by the sorted cells was examined by staining with anti-BrdU antibody

Cell-Cycle Analysis of the CD34⁻KSL and CD34⁺KSL Cells

A large body of evidence suggests that HSCs are dormant in terms of cell proliferation. However, not much is known about their cell proliferation kinetics. Because this is a crucial problem for gene therapy targeting HSCs, we isolated the CD34⁻KSL and Lin⁻ cells and performed cell cycle analysis using propidium iodide staining. As shown in Fig. 3, less than 3% of CD34⁻KSL cells were found to be in S, G_2/M phase. This result is in contrast to CD34⁺KSL cells, Jurkut cell line, or total BM cells where 26%, 56%, and 21% of the cells were in the S, G_2/M phase. These data support the previous notion that although HSCs are responsible for supplying enormous number of blood cells throughout life, they are not actively dividing. Hematopoietic progenitors appear to be responsible for the expansion of blood cells.

These data, however, do not tell us whether those CD34⁻KSL cells in the cell cycle are supplying all blood cells or, although most stem cells are participating in hemopoiesis, because they are dividing very slowly, only 3% of them are in the cell cycle at CD34⁻KSL a given time. To analyze proliferation kinetics of the HSCs, bromodeoxy uridine (BrdU) was administered daily for up to 28 days. Mice were killed at days 7, 14, and 28 and the bone marrow cells were obtained. Bone marrow cells were stained by a combination of monoclonal antibodies and the CD34⁻KSL or lineage-negative cells were FACS sorted as described [10]. Those cells were fixed and stained with anti-BrdU antibody (Becton Dickinson, San Jose, CA, USA) which was developed by FITC-conjugated goat antimouse Ig antibodies. At day 7, more than 90% of lineage-negative cells had incorporated BrdU and reached a plateau, while 30% of CD34⁻KSL cells became positive for BrdU. At day 14, 60% of CD34⁻KSL cells

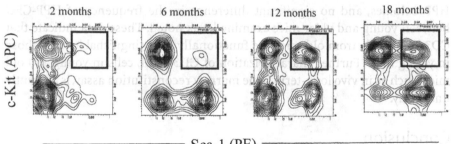

FIG. 5. Increase of CD34–KSL cells with age. Bone marrow cells obtained from 2-, 6-, 12-, and 18-month-old C57BL6 mice were stained by monoclonal antibodies for the presence of CD34⁻KSL. The expression profile of c-Kit and Sca-1 after Lin⁻ CD34⁻ gating is shown. Relative frequency of CD34⁻KSL cells as indicated by the *squares* increased with age of the mouse examined, as shown by representative data of four independent analyses

became positive, and at day 28, 70% of them were positive and the number of positive cells was still increasing (Fig. 4). These data indicate that although only 3% of HSCs are in the cell cycle at a given time, the majority of HSCs are participating in the blood supply by slowly dividing to generate progenitor cells. Such characteristics may make the use of retroviral vectors for gene therapy targeting HSCs difficult.

Age-Associated Increase of CD34⁻KSL Cells

Our previous data demonstrating long-term bone marrow reconstitution by single CD34⁻KSL cells strongly suggest that the HSCs have self-renewal capability [10]. On the other hand, proliferation capacity of a cell should be limited by the telomere length or the telomerase activity. These issues led us examine the bone marrow of aged mice. Bone marrow cells of 2-, 6-, 12-, and 18-month-old C57BL/6 mice were analyzed by FACS for the presence of CD34⁻KSL cells (Fig. 5). Surprisingly, the relative number of CD34⁻KSL cells in the bone marrow increased with age. Because the total number of bone marrow cells did not change significantly between young and old mice, this result indicated that the absolute number of CD34⁻KSL cells increased over 10 fold in 18-month-old mice compared with 2-month-old mice. To see if CD34⁻KSL cells in old mice were functionally equivalent to those in younger mice, CD34⁻KSL cells were FACS sorted from the bone marrow of 2-, 6-, 12-, and 18-month-old mice, and in vitro colony assay for HPP-CFU was performed. As described previously, in the presence of IL-3 alone, only a few colonies were formed. However, when both SCF and IL-3 were present, 80% of them formed

HPP colonies, and no significant difference in the frequency of HPP-CFU between young and older mice examined was noted. These data indicate that CD34⁻KSL cells from old mice are functionally indistinguishable from those in young mice. Further characterization of CD34⁻KSL cells in young and old mice, such as in vivo long-term bone marrow reconstitution assay, is currently underway.

Conclusion

Isolation and characterization of HSCs and progenitors as represented by CD34⁻ and CD34⁺KSL cells, respectively, have revealed significant difference in proliferation, differentiation, and self-renewing potentials. The difference is also clearly shown in the expression of various cytokine receptor genes. It is intriguing that CS34⁻KSL cells with the broadest differentiation potential express the least number of lineage-specific cytokine receptor genes. These data support the proposed notion that lineage commitment of HSCs does not depend on expression of lineage-specific cytokine receptor genes [14–16].

Despite the fact that CD34⁻KSL cells have enormous potentials as HSCs, they are too primitive and too dormant to supply sufficient blood cells to radioprotect lethally irradiated mice. On the other hand, CD34⁺KSL cells are capable of radioprotecting those mice, but only temporarily because of their lack of self-renewing capability. Thus, although even a single CD34⁻KSL cell has the potential to reconstitute the whole hematopoietic system, for successful bone marrow reconstitution, cotransplantation of either a much larger number of cells or progenitor cells such as those in the CD34⁺KSL population is required.

Acknowledgment. The authors thank Mr. Yo-hei Morita for FACS operation and Michie Ito for secretarial assistance.

References

1. Visser JW, Van BD (1990) Purification of pluripotent hemopoietic stem cells: past and present. Exp Hematol 18:248–256
2. Ogawa M, Nishikawa S, Yoshinaga K, Hayashi S, Kunisada T, Nakao J, Kina T, Sudo T, Kodama H, Nishikawa S (1993) Expression and function of c-Kit in fetal hemopoietic progenitor cells: transition from the early c-Kit independent to the late c-Kit-dependent wave of hemopoiesis in the murine embryo. Development (Camb) 117:1089–1098
3. Okada S, Nakauchi H, Nagayoshi K, Nishikawa S, Miura Y, Suda T (1992) In vivo and in vitro stem cell function of c-kit- and Sca-1-positive murine hematopoietic cells. Blood 80:3044–3050

4. Osawa M, Nakamura K, Nishi N, Takahasi N, Tokuomoto Y, Inoue H, Nakauchi H (1996) In vivo self-renewal of c-Kit+ Sca-1+ Lin(low/−) hemopoietic stem cells. J Immunol 156:3207–3214

5. Krause DS, Fackler MJ, Civin CI, May WS (1996) CD34: structure, biology, and clinical utility. Blood 87:1–13

6. Simmons DL, Satterthwaite AB, Tenen DG, Seed B (1992) Molecular cloning of a cDNA encoding CD34, a sialomucin of human hematopoietic stem cells. J Immunol 148:267–271

7. Nakamura Y, Komano H, Nakauchi H (1993) Two alternative forms of cDNA encoding CD34. Exp Hematol 21:236–242

8. Brown J, Greaves MF, Molgaard HV (1991) The gene encoding the stem cell antigen, CD34, is conserved in mouse and expressed in haemopoietic progenitor cell lines, brain, and embryonic fibroblasts. Int Immunol 3:175–184

9. Suda J, Sudo T, Ito M, Ohno N, Yamaguchi Y, Suda T (1992) Two types of murine CD34 mRNA generated by alternative splicing. Blood 79:2288–2295

10. Osawa M, Hanada K, Hamada H, Nakauchi H (1996) Long-term lymphohematopoietic reconstitution by a single CD34-low/negative hematopoietic stem cell. Science 273:242–245

11. Brady G, Iscove NN (1993) Construction of cDNA libraries from single cells. Methods Enzymol 225:611–623

12. Hu M, Krause D, Greaves M, Sharkis S, Dexter M, Heyworth C, Enver T (1997) Multi-lineage gene expression precedes commitment in the hemopoietic system. Genes Dev 11:774–785

13. Cheng J, Baumhueter S, Cacalano G, Carver MK, Thibodeaux H, Thomas R, Broxmeyer HE, Cooper S, Hague N, Moore M, Lasky LA (1996) Hematopoietic defects in mice lacking the sialomucin CD34. Blood 87:479–490

14. Pharr PN, Hankins D, Hofbauer A, Lodish HF, Longmore GD (1993) Expression of a constitutively active erythropoietin receptor in primary hematopoietic progenitors abrogates erythropoietin dependence and enhances erythroid colony-forming unit, erythroid burst-forming unit, and granulocyte/macrophage progenitor growth. Proc Natl Acad Sci USA 90:938–942

15. Dubart A, Feger F, Lacout C, Goncalves F, Vainchenker W, Dumenil D (1994) Murine pluripotent hematopoietic progenitors constitutively expressing a normal erythropoietin receptor proliferate in response to erythropoietin without preferential erythroid cell differentiation. Mol Cell Biol 14:4834–4842

16. Nishijima I, Nakahata T, Hirabayashi Y, Inoue T, Kurata H, Miyajima A, Hayashi N, Iwakura Y, Arai K, Yokota T (1995) A human GM-CSF receptor expressed in transgenic mice stimulates proliferation and differentiation of hemopoietic progenitors to all lineages in response to human GM-CSF. Mol Biol Cell 6:497–508

Ex Vivo Expansion of Human Hematopoietic Stem Cells

TATSUTOSHI NAKAHATA, MING-JIANG XU, TAKAHIRO UEDA,
SAHOKO MATSUOKA, MAMORU ITO, and KOHICHIRO TSUJI

Recent progress of embryology has provided significant insight into hematopoietic stem cell biology. There has been great interest in the ex vivo expansion of human hemopoietic stem/progenitor cells for a variety of clinical applications. I present a novel approach using soluble interleukin-6 receptor (sIL-6R) for ex vivo expansion of human hemopoietic stem/progenitor cells. Our FACS analysis revealed that most immature progenitor cells such as CFU-Mix and LTCIC were included in the $CD34^+gp130^+IL-6R^-$ population, suggesting that sIL-6R/IL-6 but not IL-6 may be potent for the ex vivo expansion of immature human progenitor cells. sIL-6R/IL-6 dramatically stimulated expansion of various progenitors including CFU-GEMM and CFU-Blast as well as $CD34^+$ cells in the presence of stem cell factor (SCF) or flt3/flk-2 ligand (FL). A combination of sIL-6R/IL-6 and SCF expanded CFU-Mix approximately 68 fold in serum-free culture by day 14. More than 100-fold expansion of total and multipotential progenitors was obtained from $CD34^+IL-6R^-$ cells but not from $CD34^+IL-6R^+$ cells in culture with sIL-6/IL-6/SCF. Addition of thrombopoietin (TPO) to culture with sIL-6R/IL-6/SCF or sIL-6/IL-6/FL significantly enhanced the expansion of immature progenitor cells. These findings suggest that gp130 in combination with Kit or flk-2/flt3 signalings may be potent for ex vivo expansion of human hematopoietic stem/progenitor cells.

Recent evidence shows that long-term repopulating hematopoietic stem cells (LTR-HSC) first occur and expand in the aorta-gonad-mesonephros (AGM) region at 10–11 days post coitum (dpc) in mice, suggesting that the AGM region at this stage provides a microenvironment suitable for development for hematopoietic stem cells. We established a novel stromal cell line, AGM-S3, from the AGM region at 10.5 dpc, which supported for 6 weeks the

Department of Clinical Oncology, Institute of Medical Science, University of Tokyo, 4-6-1 Shiroganedai, Minato-ku, Tokyo 108-8639, Japan

generation of human multipotential progenitors from CD34$^+$CD38$^-$ primitive hematopoietic cells. Human LTR-HSC with the potential to reconstitute hematopoiesis in NOD/SCID mice were maintained on AGM-S3 cells for at least 4 weeks. Flow cytometric analysis showed that CD13, vascular cellular adhesion molecule-1 and Sca-1 were expressed on AGM-S3 cells. Because SCF, IL-6, and oncostatin M, but not IL-3, IL-11, leukemia inhibitory factor, granulocyte colony-stimulating factor, granulocyte-macrophage colony-stimulating factor, TPO, and FL were detected in reverse transcription polymerase chain reaction analysis of AGM-S3 cells, the cells seem to express some molecules that act on human hematopoietic stem cells. This cell line is expected to elucidate molecular mechanisms regulating early hematopoiesis and thus to pave the way for developing strategies for expansion of human LTR-HSC.

Notch in Hematopoiesis: Cell Fate Decisions and Self-Renewal of Progenitors

Laurie A. Milner[1] and Anna Bigas[2]

Summary. The Notch family comprises a group of highly conserved cell-surface receptors that mediate cell fate decisions in many developmental processes, from worms and flies to mammalian systems. Cell–cell signaling through Notch permits multipotent progenitors in the same environmental context to respond differently to developmental signals, facilitating the establishment of distinct cell types. Over the past several years major advances have been made in defining the molecular mechanisms involved in Notch signaling, understanding the complex interactions of Notch with other signaling pathways, and establishing the central role of Notch in mammalian development. Here we present evidence supporting a conserved role for Notch in hematopoietic regulation. Notch1 and Notch2 are both expressed by hematopoietic progenitors and the Notch ligand, Jagged1, is expressed by a subset of bone marrow stromal cells. Using 32D myeloid progenitors, we show that activation of Notch1 by Jagged1 inhibits G-CSF-induced granulocytic differentiation and permits the maintenance of undifferentiated cells. We also demonstrate that Notch1 and Notch2 have distinct intracellular activities that permit modulation of myeloid differentiation specifically in response to G-CSF or GM-CSF. We speculate that Notch plays a central role in the regulation of cell fate determination and self-renewal of progenitors during hematopoiesis.

Key words. Notch, Myeloid differentiation, Progenitor self-renewal, Cell fate decisions

[1] The Fred Hutchinson Cancer Research Center, 1100 Fairview Ave N., Seattle, WA 98109-1024, and the Department of Pediatrics, University of Washington School of Medicine, Seattle, WA 98109, USA
[2] Institut de Recerca Oncologica, Hospital Duran y Reynals, Barcelona, Spain

Introduction

Hematopoiesis is a continuous developmental process in which pluripotent stem cells and their progeny make sequential cell fate decisions, producing mature blood cells of the various lineages. The constant generation of appropriate numbers and types of mature cells, as well as the maintenance of multipotent progenitors, requires a complex regulatory network. Although much progress has been made in delineating these pathways, several key aspects of hematopoiesis, including the mechanisms responsible for cell fate specification and self-renewal of progenitors, remain incompletely understood and controversial [1–7].

Members of the Notch family play critical roles in cell fate determination and maintenance of progenitors in a variety of developmental processes [8–13]. Notch functions both as a receptor in cell–cell interactions and as a transcriptional regulator, mediating signal transduction directly from the cell surface to the nucleus, and permitting cells to directly influence gene expression in their neighbors [14–17]. The unique role of Notch as a regulator of cell fate decisions in so many systems stems from its capacity to engage in a variety of molecular interactions, integrating inductive signals to influence gene expression in a way that reflects the precise developmental context of a particular cell. Notch interacts with a host of factors that are important in hematopoietic regulation, including components of cytokine pathways, developmental pathways, and transcriptional regulators. The integration of Notch with these other regulatory pathways is likely to be an important component of hematopoietic regulation.

Conservation of the Notch Signaling Pathway

The broad evolutionary conservation of Notch function is reflected in the structural conservation of Notch proteins and components of the Notch signal transduction pathway. Figure 1a depicts the general structure of Notch, noting domains of functional significance. The Notch extracellular domain contains a variable number of tandem epidermal growth factor (EGF) -like repeats and three Lin/Notch repeats (LNR), which function in specific ligand binding and Notch activation. The Notch intracellular domain includes six cdc10/ankyrin repeats; this is the most highly conserved region and is essential for Notch signal transduction. More recently described regions include the RAM domain, which binds CSL (CBF1/Su(H)/Lag-1) effector molecules [18–21] and the Notch cytokine response (NCR) region associated with cytokine-specific effects of Notch1 and 2 on myeloid differentiation [22].

Notch ligands (Fig. 1b) are also transmembrane proteins with an extracellular domain consisting of EGF repeats. They also contain a characteristic

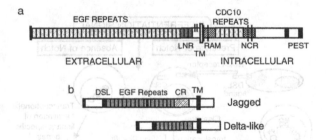

FIG. 1a,b. General structure of Notch receptors and DSL ligands, illustrating conserved functional domains. a The extracellular domain of Notch consists of 29–36 tandem EGF-like repeats and 3 Lin/Notch repeats (LNR), which are involved in DSL ligand binding and activation of Notch. The intracellular domain includes 6 cdc10 repeats, which mediate protein–protein interactions that are crucial for Notch activity. The RAM domain binds effector CSL proteins, and the Notch cytokine receptor (NCR) region is associated with cytokine-specific effects of Notch1 and 2. b Like Notch receptors, the extracellular domains of DSL ligands contain tandem EGF repeats. The DSL domain is unique to Notch ligands, and this motif is thought to be crucial for specific binding to Notch receptors. Jagged (Serrate) ligands also contain a cysteine-rich (CR) region not present in Delta (Delta-like) ligands. The intracellular domains of DSL ligands are short and have little sequence homology

motif, the DSL (Delta/Serrate/Lag-2) domain, which is unique to Notch ligands [23,24]. The two general classes of mammalian Notch ligands are Jagged (or Serrate) and Delta (or Delta-like).

The prevailing model for Notch signal transduction is illustrated in Fig. 2. In addition to Notch receptors, integral components of the Notch pathway include ligands homologous to *Drosophila* Delta and Serrate and *C. elegans* Lag-2 (DSL proteins) and intracellular effector molecules homologous to Suppressor of Hairless (CSL proteins) and Enhancer of split (E[spl]) [10–12,14]. Activation of Notch occurs through binding of Notch to a DSL ligand, resulting in proteolytic cleavage and translocation of the Notch intracellular domain to the nucleus [25–27]. In the nucleus, Notch participates in transcriptional regulation that results in the suppression of lineage-specific genes and inhibition of differentiation. In the absence of Notch activation, an equivalent cell exposed to the same differentiative signal will differentiate according to a molecular program that reflects a defined pattern of lineage-specific gene expression and results in the production of a mature cell.

Intercellular Signaling Through Notch: Homotypic and Heterotypic Interactions

Notch signaling can occur either among a group of equivalent cells (homotypic interactions) or between nonequivalent cells (heterotypic interactions),

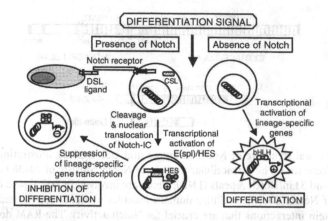

FIG. 2. Signal transduction through the Notch pathway. In the presence of a specific differentiation signal, activation of Notch in response to ligand binding results in proteolytic cleavage and release of the Notch intracellular domain. Intracellular Notch translocates to the nucleus, where it participates in transcriptional activation of negative transcriptional regulators, such as E(spl)/HES. The transcription factors encoded by E(spl)/HES in turn suppress transcription of lineage-specific genes, thereby inhibiting cellular differentiation. An equivalent cell, in the absence of Notch activation (*right*), will respond to the differentiation signal by activating transcription of lineage-specific genes, permitting differentiation along the induced pathway

both of which are essential during development [8,9,11,15]. In the hematopoietic system, Notch is expressed by hematopoietic progenitors, and DSL ligands are expressed by both stromal and hematopoietic cells, suggesting the importance of both types of interactions. In lateral signaling (Fig. 3), among a group of equipotent progenitors that express both Notch and DSL ligand, a signaling bias is established based on minor differences in Notch and ligand expression. When exposed to the inductive signal A, a limited number (those expressing less Notch/more DSL ligand) adopt the specific fate A, while adjacent cells (expressing more Notch) are inhibited from differentiating. These progenitors remain competent to respond to subsequent signals, but again establish a signaling bias that limits the number adopting fate B in response to signal B. The same occurs for remaining progenitors exposed to signals C and D. The result is the generation of cells of multiple distinct lineages as well as maintenance of multipotent progenitors.

Notch in Hematopoiesis

We have hypothesized that Notch functions at differentiation branch points throughout hematopoiesis to influence lineage specification and self-renewal

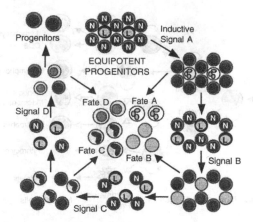

FIG. 3. The relative levels of Notch expression determine sequential cell fates. Within a group of equipotent progenitors, cells expressing more DSL ligand (L) and less Notch escape from the inhibitory effects of Notch and respond to the inductive *Signal A* by adopting the primary cell *Fate A*; adjacent cells that express relatively more Notch (N) are inhibited from adopting Fate A, but remain competent to respond to subsequent signals. Among these remaining progenitors, differential expression of Notch and DSL ligand again determines which cells will respond to the inductive *Signal B*: those expressing more ligand adopt *Fate B*, while those expressing more Notch are again inhibited from differentiating. Again, the remaining progenitors establish differential Notch–DSL ligand expression, permitting only some cells to respond to the subsequent *Signals C* and *D*. Thus, from a group of originally equipotent progenitors, cells of multiple distinct lineages are established, and some cells are maintained as uncommitted progenitors

of progenitors (Fig. 4). Since the identification of the human *Notch1* gene through its association with T-cell leukemias [28] and the demonstration that *Notch1* is expressed by normal bone marrow hematopoietic precursors [29], a significant amount of evidence has emerged to support this hypothesis (for review, see [30]). In both the bone marrow and thymus, *Notch1* is expressed at relatively high levels in the least mature progenitors and at lower levels in more mature cells, an expression pattern consistent with the view that Notch maintains cells, in a less differentiated state [29,31,32]. The expression of Notch ligands by bone marrow and fetal liver stromal cells and by thymic epithelial cells further indicates that Notch signaling occurs in the hematopoietic and lymphopoietic microenvironments [33–37].

Notch Function in Myeloid Differentiation

We have used the 32D myeloid progenitor cell line as a model system to study the effects of Notch signaling on myeloid differentiation. The 32D Cl3 cell line was established from normal mouse bone marrow cultures and represents a population of hematopoietic progenitors at relatively early stages of myeloid

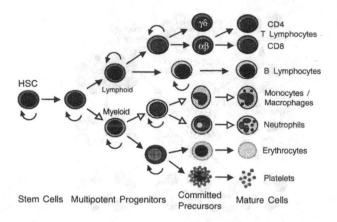

FIG. 4. Cell fate decisions in hematopoiesis. During hematopoiesis, pluripotent stem cells (*HSC*) and multipotent progenitors make sequential cell fate decisions to produce mature blood cells of the various lineages (*solid arrows*). In addition, some progenitors must self-renew to maintain a pool of uncommitted progenitors. We speculate that Notch influences lineage decisions and self-renewal of progenitors at each branch point during hematopoiesis. In this chapter, we present evidence for Notch function in myeloid differentiation (*open arrows*)

commitment [38]. 32D cells proliferate as undifferentiated blasts in the presence of IL-3, but can be induced to differentiate in the presence of other cytokines, including granulocyte colony-stimulating factor (G-CSF) [39] and granulocyte-macrophage colony-stimulating factor (GM-CSF) [40]. In the studies presented here, we used retroviral transduction to express different forms of Notch1 and 2 in 32D cells, and evaluated cytokine-induced differentiation compared to control 32D cells. In initial studies we found that expression of an activated intracellular Notch1 molecule inhibited differentiation in response to G-CSF and permitted the continued proliferation of undifferentiated cells, findings consistent with the effects of constitutive Notch activity in other systems [41].

Activation of Notch1 by Jagged1 inhibits differentiation and permits the maintenance of undifferentiated 32D progenitors

Signaling through the Notch pathway in the marrow microenvironment is suggested by the expression of *Jagged1* by bone marrow (BM) stromal cells and *Notch1* by immature hematopoietic progenitors. *Jagged1* is expressed by the stromal cell line HS-27a, but not by other stromal lines, including HS-5 and HS-23 [33]. HS-27a promotes cobblestone area formation in long-term marrow cultures and supports the maintenance and proliferation of hematopoietic progenitors [42], properties that could be attributed to Jagged–Notch signaling. To test this hypothesis, we evaluated the effects of

HS-27a on G-CSF-induced differentiation of 32D cells expressing full-length Notch1 (FLN). We found that HS-27a specifically inhibited differentiation of FLN1 32D cells, but not of control (LXSN-transduced) 32D cells (Fig. 5a). Furthermore, FLN 32D cells differentiated normally in the presence of G-CSF alone or when cocultured with control cell lines (HS-5 and HS-23; data not shown). Control 32D cells differentiated in the presence of HS-27a

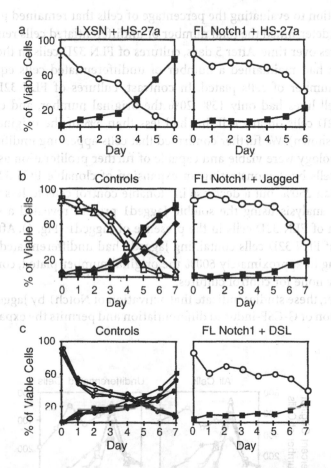

FIG. 5a–c. Activation of Notch1 by Jagged1 inhibits granulocytic differentiation of 32D cells. Graphs illustrate the percentages of viable cells that remained undifferentiated (*open symbols*) or had become mature granulocytes (*closed symbols*) after successive days in culture with G-CSF. Each graph represents three separate experiments, each evaluating three individual clones for each construct. Error bars are omitted for the sake of clarity. **a** The effects of the Jagged1-expressing HS-27a cell line on G-CSF-induced differentiation of 32D cells expressing the full-length (FL) form of Notch1 are compared to 32D cells expressing a control retroviral construct (LXSN). **b** The effects of a soluble extracellular Jagged1 protein (produced in COS cells, and provided as COS supernatant) are compared to control COS supernatant on differentiation of FL Notch1 expressing and control 32D cells. **c** The effects of a small Jagged1 DSL peptide are compared to two control peptides on differentiation of FL Notch1 expressing and control 32D cells

as well as control cell lines [33]. To confirm that the effects of HS-27a on FLN1 32D differentiation could be attributed to signaling between Jagged1 and Notch1, we evaluated the effects of two soluble forms of Jagged1 in this assay. In these studies we found that a soluble extracellular Jagged1 protein as well as a small peptide corresponding to part of the unique DSL domain had effects comparable to the HS-27a cell line (Fig. 5b,c).

In addition to evaluating the percentage of cells that remained undifferentiated, we determined the total number of undifferentiated cells remaining in the cultures over time. After 5 days, cultures of FLN 32D cells in the presence of HS-27a had maintained a number of undifferentiated cells equal to the original number of cells plated. In contrast, cultures of FLN 32D cells on control cell lines had only 15%–20% the original number, and cultures of control 32D cells on any cell line had less than 5% of the original number (data not shown). We further confirmed that cells appearing undifferentiated by morphology were viable and capable of further proliferation as undifferentiated cells by demonstrating an expansion of clonable FLN 32D cells of greater than 250%, but a decrease in clonable control cells to less than 10%. A similar analysis using the soluble Jagged1 protein revealed a significant expansion of FLN 32D cells in the presence of Jagged1 (Fig. 6). After 5 days, cultures of FLN 32D cells containing Jagged1 had undifferentiated cells corresponding to approximately 600% the original number plated, compared to essentially none for control cultures.

Together, these studies indicate that activation of Notch1 by Jagged1 results in inhibition of G-CSF-induced differentiation and permits the expansion and

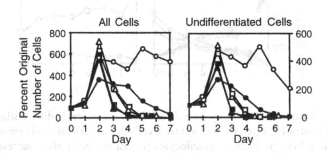

FIG. 6. Proliferation and maintenance of undifferentiated FLN 32D cells in the presence of a soluble extracellular Jagged1 protein. The relative number of all cells (*left*) and undifferentiated cells (*right*) remaining in cultures over time are depicted as the percent of the original number of cells plated. *Open circles* represent cultures of FL Notch 32D cells in the presence of Jagged1 protein; other symbols represent control 32D cells in the presence of Jagged1 and control supernatant, and FL Notch 32D cells in the presence of control supernatant

maintenance of 32D myeloid progenitors. The effects of Notch1 activation through ligand binding were comparable to those observed with expression of a constitutively activated form of Notch1 in this system [41], further validating the use of activated intracellular forms of Notch as a model for Notch activity in 32D cells. However, the effects of Notch activation by ligand binding were more transient than the effects of constitutively activated Notch expression, suggesting the presence of normal regulatory mechanisms in addition to an intact Notch signaling pathway in 32D cells.

Cytokine-Specific Effects of Notch1 and 2: The Notch Cytokine Response (NCR) Region

Notch1 and 2 are both expressed in myeloid progenitors, raising the possibility that they have distinct functions in myelopoiesis. In studies to address this question, we evaluated the capacity of 32D cells expressing activated intracellular forms of Notch1 or Notch2 to differentiate in response to G-CSF or GM-CSF. Control 32D cells differentiate in response to either G-CSF or GM-CSF (Fig. 7); cells expressing activated Notch1 or Notch2 are inhibited from differentiating in a cytokine-specific manner: Notch1 inhibits differentiation in response to G-CSF, whereas Notch2 inhibits differentiation in response to GM-CSF (Fig. 7). These findings suggest that Notch1 and 2 have distinct intracellular molecular interactions that permit different effects in the same cell type.

To define which part of the Notch molecule is responsible for cytokine specificity, we generated chimeric Notch1/Notch2 molecules and evaluated their effects on 32D differentiation. The chimeric molecules included the Notch1 or 2 cdc10 domain and the reciprocal C-terminal flanking region. These studies revealed that the flanking region, which we have termed the Notch Cytokine Response (NCR) region, confers cytokine specificity on the Notch1 and Notch2 molecules. As shown in Fig. 7, the chimeric molecule containing the Notch1 NCR (Notch2CDC/Notch1NCR) had effects comparable to Notch1 (inhibited differentiation specifically in response to G-CSF), and the molecule containing the Notch2 NCR (Notch1CDC/Notch2NCR) had effects comparable to Notch2 (inhibited differentiation in response to GM-CSF). We also found that the Notch1 and 2 NCR regions conferred differences in subcellular localization and electrophoretic mobility, suggesting that differences in posttranslational modification of the NCR region may dictate distinct interactions of Notch1 and 2 with components of the G-CSF and GM-CSF signaling pathways [22]. Thus, these studies elucidate a potentially important link between Notch and cytokine signaling pathways in the regulation of hematopoietic differentiation.

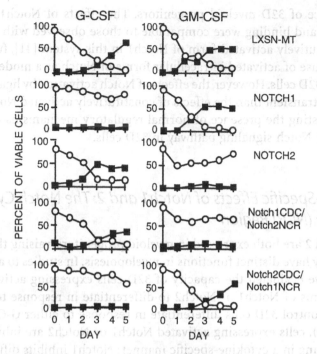

FIG. 7. Effects of activated forms of Notch1 and Notch2 on G-CSF- and GM-CSF-induced granulocytic differentiation of 32D cells. Individual clones transduced with a control myc tag (LXSN-MT) retroviral construct, activated forms of Notch1 or Notch2, or chimeric Notch1/Notch2 molecules were evaluated for proliferation and differentiation in the presence of G-CSF or GM-CSF. Graphs show the relative percentage of viable differentiated (*closed squares*) and undifferentiated (*open circles*) cells present in the culture after successive days. The same three independent clones for each construct were used for both G-CSF and GM-CSF differentiation

Conclusions: A Model for Notch Function in Hematopoiesis

The studies presented here provide substantiating evidence for a role for Notch in hematopoietic cell fate decisions, and further suggest that different Notch molecules have distinct functions in the hematopoietic system. When considered together with the known role of Notch in other systems, our findings indicate that under appropriate conditions activation of Notch will delay differentiation and permit self-renewal of hematopoietic stem and progenitor cells. However, it is also clear that the effects of Notch depend on the precise developmental context, including cell–cell interactions and cytokine stimuli.

The model in Fig. 8 depicts selected interactions that illustrate how Notch may function to influence myeloid cell fate decisions and to maintain multipotent progenitors at various stages of maturation. Although the model is based on known features of Notch signaling, details of Notch function in hematopoiesis are lacking for all but a very few aspects. The model predicts that in the marrow microenvironment Notch signaling occurs between stromal cells and hematopoietic progenitors, as well as between equivalent and nonequivalent hematopoetic cells. Hematopoietic progenitors express

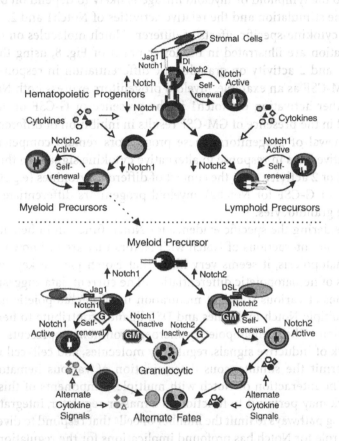

FIG. 8. A model for Notch function in hematopoiesis. The *top panel* illustrates potential effects of Notch signaling on differerentiation and self-renewal of immature hematopoietic progenitors. Hematopoietic progenitors express multiple Notch molecules (depicted as Notch1 and 2) and DSL ligands. Stromal cells express DSL ligands, including Jagged and Delta. Notch signaling occurs between stromal cells and hematopoietic progenitors, and between equivalent or nonequivalent hematopoietic cells. In the context of various cytokines, progenitors are induced to differentiate. Notch signaling regulates the response of progenitors to cytokine stimulation, permitting some to differentiate and others to self-renew. Cells expressing more Notch are inhibited from differentiating, and thus maintain a pool of uncommitted progenitors. Cells expressing less Notch escape from the Notch signal and undergo the next step in differentiation. Myeloid progenitors (*lower panel*) also express both Notch1 and 2, and the effects on differentiation are cytokine specific. The effects of granulocytic differentiation in response to G-CSF and GM-CSF described in the text are used for illustration purposes

multiple Notch molecules and Notch ligands. Stromal cells express Notch ligands, including Jagged and Delta. In the context of cytokine stimulation, some progenitors differentiate while others self-renew, a decision likely mediated by Notch signaling. We would predict that cells expressing less Notch will differentiate, while those expressing more Notch will be inhibited from differentiating, thus maintaining a pool of uncommitted progenitors. Commitment to the lymphoid or myeloid lineage is likely to depend on both specific cytokine stimulation and the relative activities of Notch1 and 2.

The cytokine-specific effects of different Notch molecules on myeloid differentiation are illustrated in the lower panel of Fig. 8, using the effects of Notch1 and 2 activity on granulocytic differentiation in response to G-CSF and GM-CSF as an example. Myeloid progenitors express both Notch1 and 2 and either activation of Notch1 in the presence of G-CSF or activation of Notch2 in the presence of GM-CSF results in inhibition of differentiation and self-renewal of progenitors. These progenitors remain competent to adopt alternative fates in response to alternative cytokine signals. In the absence of Notch1 or 2 activity, or in the context of different cytokines (e.g., GM-CSF for Notch1 or G-CSF for Notch2), myeloid progenitors differentiate to produce mature granulocytes.

Considering the specific evidence for Notch function in hematopoiesis, as well as the interactions of Notch with factors that are of known importance in hematopoiesis, it seems very likely that Notch plays a key role in many aspects of hematopoietic differentiation. The current data suggest that Notch functions at various stages of maturation in all hematopoietic lineages, and that multiple Notch molecules and DSL ligands contribute to hematopoietic regulation. The hematopoietic microenvironment represents a complex network of inductive signals, regulatory molecules, and cell–cell interactions that permit the simultaneous determination of various hematopoietic cell fates. The interaction of Notch with multiple components of this regulatory network may permit it to function as a master regulator, integrating various signaling pathways to limit the number of cells that respond to diverse signals. Such a role for Notch has profound implications for the regulation of cell fate determination and self-renewal of hematopoietic progenitors.

Stem Cell Expansion

Because Notch primarily functions to delay differentiation and permit self-renewal of multipotent progenitors, it may be possible to exploit Notch signaling for in vitro expansion of hematopoietic stem/progenitor cells. Preliminary studies adding DSL ligands to primary hematopoietic cultures indicate that this may be a successful strategy [34,35]. However, a more complete understanding of Notch signaling in hematopoiesis is clearly needed to determine the scope of effects and establish optimal conditions.

Gene Therapy

Notch signaling may also prove to be be useful for genetic manipulation of hematopoietic stem cells and thus generally applicable to gene therapy. Under certain conditions, Notch activity induces mitotic activity while also preventing differentiation [43]. Therefore, activation of Notch may facilitate retroviral gene transfer into stem cells by promoting cell cycling without inducing differentiation. Following retroviral transduction, it may also be possible to use Notch signaling to expand transduced cells in vitro or to improve long-term survival of transduced stem cells in vivo.

References

1. Ogawa M (1993) Differentiation and proliferation of hematopoietic stem cells. Blood 81:2844–2853
2. Orkin SH (1995) Hematopoiesis: how does it happen? Curr Biol 7:870–877
3. Shivdasani RA, Orkin SH (1996) The transcriptional control of hematopoiesis. Blood 87:4025–4039
4. Morrison SJ, Uchida N, Weissman IL (1995) The biology of hematopoietic stem cells. Annu Rev Cell Dev Biol 11:35–71
5. Morrison SJ, Shah NM, Anderson DJ (1997) Regulatory mechanisms in stem cell biology. Cell 88:287–298
6. Cross MA, Enver T (1997) The lineage commitment of haemopoietic progenitor cells. Curr Opin Genet Dev 7:609–613
7. Metcalf D, Enver T, Heyworth CM, Dexter TM (1998) Controversies in hematology: growth factors and hematopoietic cell fate. Blood 92:345–352
8. Simpson P (1997) Notch signalling in development: on equivalence groups and asymmetric developmental potential. Curr Opin Genet Dev 7:537–542
9. Kimble J, Simpson P (1997) The Lin-12/Notch signaling pathway and its regulation. Annu Rev Cell Dev Biol 13:333–361
10. Fleming RJ, Purcell K, Artavanis-Tsakonas S (1997) The NOTCH receptor and its ligands. Trends Cell Biol 7:437–441
11. Greenwald I (1998) LIN-12/Notch signaling: lessons from worms and flies. Genes Dev 12:1751–1762
12. Egan SE, St. Pierre B, Leow CC (1998) Notch receptors, partners and regulators: from conserved domains to powerful functions. Curr Top Microbiol Immunol 228:273–324
13. Weinmaster G (1998) Notch signaling: direct or what? Curr Opin Genet Dev 8:436–442
14. Artavanis-Tsakonas S, Matsuno K, Fortini ME (1995) Notch signaling. Science 268:225–232
15. Kopan R, Turner DL (1996) The Notch pathway: democracy and aristocracy in the selection of cell fate. Curr Opin Neurobiol 6:594–601
16. Honjo T (1996) The shortest path from the surface to the nucleus: RBP-Jκ/Su(H) transcription factor. Genes Cells 1:1–9
17. Lewis J (1998) A short cut to the nucleus. Nature (Lond) 393:304–305
18. Tamura K, Taniguchi Y, Minoguchi S, Sakai T, Tun T, Furukawa T, Honjo T (1995) Physical interaction between a novel domain of the receptor Notch and the transcription factor RBP-Jκ/Su(H). Curr Biol 5:1416–1423

19. Roehl H, Bosenberg M, Blelloch R, Kimble J (1996) Roles of the RAM and ANK domains in signaling by the *C. elegans* GLP-1 receptor. EMBO J 15:7002–7012
20. Hsieh JJ, Henkel T, Salmon P, Robey E, Peterson MG, Hayward SD (1996) Truncated mammalian Notch1 activates CBF1/RBPJκ-repressed genes by a mechanism resembling that of epstein-barr virus EBNA2. Mol Cell Biol 16:952–959
21. Kato H, Taniguchi Y, Kurooka H, Minoguchi S, Sakai T, Nomura-Okazaki S, Tamura K, Honjo T (1997) Involvement of RBP-J in biological functions of mouse Notch1 and its derivatives. Development (Camb) 124:4133–4141
22. Bigas A, Martin DIK, Milner LA (1998) Notch1 and Notch2 inhibit myeloid differentiation in response to different cytokines. Mol Cell Biol 18:2324–2333
23. Muskavitch MAT (1994) Delta-notch signaling and *Drosophila* cell fate choice. Dev Biol 166:415–430
24. Fitzgerald K, Greenwald I (1995) Interchangeability of *Caenorhabditis elegans* DSL proteins and intrinsic signalling activity of their extracellular domains in vivo. Development (Camb) 121:4275–4282
25. Blaumueller CM, Qi H, Zagouras P, Artavanis-Tsakonas S (1997) Intracellular cleavage of Notch leads to a heterodimeric receptor on the plasma membrane. Cell 90:281–291
26. Kopan R, Schroeter EH, Weintraub H, Nye J (1996) Signal transduction by activated mNotch: importance of proteolytic processing and its regulation by the extracellular domain. Proc Natl Acad Sci USA 93:1683–1688
27. Schroeter EH, Kisslinger JA, Kopan R (1998) Notch-1 signalling requires ligand-induced proteolytic release of intracellular domain. Nature (Lond) 393:382–386
28. Ellisen LW, Bird J, West DC, Soreng AL, Reynolds TC, Smith SD, Sklar J (1991) TAN-1, the human homolog of the *Drosophila* Notch gene, is broken by chromosomal translocations in T lymphoblastic neoplasms. Cell 66:649–661
29. Milner LA, Kopan R, Martin DIK, Bernstein ID (1994) A human homologue of the *Drosophila* developmental gene, Notch, is expressed in CD34$^+$ hematopoietic precursors. Blood 83:2057–2062
30. Milner LA, Bigas A (1999) Notch as a mediator of cell fate determination in hematopoiesis: evidence and speculation. Blood 93:2431–2448
31. Hasserjian RP, Aster JC, Davi F, Weinberg DS, Sklar J (1996) Modulated expression of Notch1 during thymocyte development. Blood 88:970–976
32. Robey E, Chang D, Itano A, Cado D, Alexander H, Lans D, Weinmaster G, Salmon P (1996) An activated form of Notch influences the choice between CD4 and CD8 T cell lineages. Cell 87:483–492
33. Li L, Milner LA, Deng Y, Iwata M, Banta A, Graf L, Marcovina S, Friedman C, Trask B, Hood L, Torok-Storb B (1998) The human homolog of rat Jagged, hJagged1, is expressed by marrow stroma and inhibits differentiation of 32D cells through interaction with Notch1. Immunity 8:43–55.
34. Varnum-Finney B, Purton LE, Yu M, Brashem-Stein C, Flowers D, Staats S, Moore KA, Le Roux I, Mann R, Gray G, Artavanis-Tsakonas S, Bernstein ID (1998) The Notch ligand, Jagged-1, influences the development of primitive hematopoietic precursor cells. Blood 91:4084–4091
35. Jones P, May G, Healy L, Brown J, Hoyne G, Delassus S, Enver T (1998) Stromal expression of Jagged1 promotes colony formation by fetal hematopoietic progenitor cells. Blood 92:1505–1511
36. Shawber C, Boulter J, Lindsell CE, Weinmaster G (1996) Jagged2: a serrate-like gene expressed during rat embryogenesis. Dev Biol 180:370–376

37. Luo B, Aster JC, Hasserjian RP, Kuo F, Sklar J (1997) Isolation and functional analysis of a cDNA for human Jagged2, a gene encoding a ligand for the Notch1 receptor. Mol Cell Biol 17:6057–6067
38. Greenberger JS, Sakakeeny MA, Humphries RK, Eaves CJ, Eckner RJ (1983) Demonstration of permanent factor-dependent multipotential (erythroid/neutrophil/-basophil) hematopoietic progenitor cell lines. Proc Natl Acad Sci USA 80:2931–2935
39. Valtieri M, Tweardy DJ, Caracciolo D, Johnson K, Mavilio F, Altmann S, Santoli D, Rovera G (1987) Cytokine-dependent granulocytic differentiation: regulation of pro-liferative and differentiative responses in a murine progenitor cell line. J Immunol 138:3829–3835
40. Kreider BL, Phillips PD, Prystowsky MB, Shirsat N, Pierce JH, Tushinski R, Rovera G (1990) Induction of the granulocyte-macrophage colony-stimulating factor (CSF) receptor by granulocyte CSF increases the differentiative options of a murine hematopoietic progenitor cell. Mol Cell Biol 10:4846–4853
41. Milner LA, Bigas A, Kopan, Brashem-Stein C, Black M, Bernstein ID, Martin DIK (1996) Inhibition of granulocytic differentiation by mNotch1. Proc Natl Acad Sci USA 93:13014–13019
42. Roecklein BA, Torok-Storb B (1995) Functionally distinct human marrow stromal cell lines immortalized by transduction with the human papilloma virus E6/E7 genes. Blood 85:1005
43. Go MJ, Eastman DS, Artavanis-Tsakonas S (1998) Cell proliferation control by Notch signaling in *Drosophila* development. Development (Camb) 125:2031–2040

Molecular Genetics of Acute Promyelocytic Leukemia: A Rationale for "Transcription Therapy" for Cancer

PIER PAOLO PANDOLFI

Summary. The last few years have been crucial for the elucidation of the molecular mechanisms underlying the pathogenesis of human leukemia [1]. In particular, more than 80% of myeloid leukemias have been attributed to, or associated with, one or more specific molecular lesions. In the vast majority of cases these molecular events are chromosomal translocations that rearrange the regulatory and coding regions of a variety of genes which encode transcription factors. These proteins can interfere with the normal transduction, at the transcription level, of pivotal cellular processes such as growth, differentiation, and survival. The identification of these lesions renders it possible to reclassify myeloid leukemias according to new "molecular" criteria, and to develop new diagnostic and prognostic tools. Furthermore, the understanding of the aberrant transcriptional mechanisms underlying leukemia pathogenesis allows the development of new therapeutic approaches. In particular, the recent elucidation of the molecular mechanisms underlying the pathogenesis of acute promyelocytic leukemia has allowed us to propose and exploit what we regard as a new concept for the treatment of cancer, which we refer to as "transcription therapy."

Key words. Transcription factors, Transcription therapy, HDAC inhibitors, Differentiation, Cancer

Department of Human Genetics and Molecular Biology Program, Memorial Sloan-Kettering Cancer Center, Sloan-Kettering Division, Graduate School of Medical Sciences, Cornell University, 1275 York Avenue, New York, NY 10021, USA

Acute Promyelocytic Leukemia: A Paradigm for Aberrant Transcription in Cancer

Acute promyelocytic leukemia (APL) is characterized by the clonal expansion of malignant myeloid cells blocked at the promyelocytic stage of hemopoietic development [2,3 and references therein]. APL is associated with reciprocal chromosomal translocations always involving the retinoic acid receptor-α (*RARα*) gene on chromosome 17 [4,5]. In most APL patients, the translocation involves chromosome 15 and chromosome 17 [t(15;17) (q22;21)] [4,5]. The breakpoints on chromosome 15 cluster within a locus originally called *myl* and now named *PML* (for promyelocytic leukemia gene). In a few cases, the translocation involves chromosome 11 instead of chromosome 15, t(15;17) (q23;21), and a gene named promyelocytic leukemia zinc finger (*PLZF*).

In only two cases so far described, the translocation involves chromosome 5 and the *Nucleophosmin* gene (*NPM*), t(5;17) (q32;q12). The common feature of the three translocations is the involvement of chromosome 17 and the *RARα* locus. As a consequence of the translocation, two fusion genes are produced, and at least in the t(15;17) and t(11;17) cases, both of these are transcribed into their respective fusion proteins (for brevity referred to hereafter as X-RARα and RARα-X fusion proteins). Very recently a new case of APL has been reported where once again *RARα* translocates to chromosome 11. In this newly described translocation, t(11;17) (q13;q21), the breakpoint on chromosome 11 has been mapped to the *NuMA* gene [4].

PML is a member of the ring finger family of proteins [4–7]. The *PLZF* gene encodes a transcription factor with transcriptional repressive activity, and is a member of the POK (POZ and *Krüppel*) family of proteins, which share an N-terminal POZ motif and a C-terminal DNA-binding domain made by *Krüppel*-like C2-H2 zinc fingers [4,5,8,9]. Although structurally unrelated, PML and PLZF heterodimerize and can colocalize in the nucleus onto structures known as nuclear bodies [4,5,10].

RARs are members of the superfamily of nuclear hormone receptors, which act as RA-inducible transcriptional activators in their heterodimeric form with retinoid-X receptors (RXRs), a second class of nuclear retinoid receptors [11]. In the absence of RA, RAR/RXR heterodimers can repress transcription through histone deacetylation by recruiting nuclear receptor corepressors (N-CoR or SMRT), Sin3A or Sin3B, which in turn form complexes with histone deacetylases (HDAC-1 or -2), thereby resulting in nucleosome assembly and transcriptional repression [11,12]. RA causes the dissociation of the corepressors complex and the recruitment of transcriptional coactivators to the RAR/RXR complex, thus resulting in the activation of gene expression which, in turn, can induce terminal differentiation and growth arrest of cells of various histological origins including normal myeloid hemopoietic cells [11–13].

The X-$RAR\alpha$ and $RAR\alpha$-X fusion genes generated by the reciprocal translocation in APL encode for structurally different X-RARα and RARα-X products, coexpressed in the leukemic blast, which differ in their X-portions, but are identical in their RARα portion, and that can therefore be considered as RARα mutants [4,5].

APL blasts are exquisitely sensitive to the differentiating action of retinoic acid (RA) [2,3,11,13]. RA can overcome the block of maturation at the promyelocytic stage and induce the malignant cells to terminally mature into granulocytes. This therapeutic approach is conceptually new in that it does not involve chemical or physical agents that eradicate the tumor by "killing" the neoplastic cells, but rather reprograms these cells to differentiate normally. APL has become, for this reason, the paradigm for "cancer differentiation therapy." However, although effective, treatment with RA in APL patients induces disease remission transiently and relapse is inevitable. Furthermore, APL associated with chromosomal translocations between the $RAR\alpha$ and the $PLZF$ genes ($PLZF$-$RAR\alpha$) shows a distinctly worse prognosis with poor response to chemotherapy and little or no response to treatment with RA, thus defining a new APL syndrome [14].

In these patients, resistance to RA could be conferred by the PLZF-RARα protein itself, or by the RARα-PLZF fusion protein, which could function as an aberrant transcription factor because it retains part of the PLZF DNA-binding domain and still binds to PLZF DNA-binding sites [9]. PML-RARα and PLZF-RARα are both able to bind to retinoic acid response elements (RARE) as homodimers and can form multimeric complexes with RXRs [15,16]. Therefore, X-RARα proteins are thought to interfere with the normal RAR/RXR-RA pathway in a dominant-negative manner through their ability to complex with RXR or through their altered DNA-binding and transcriptional activities [15,16–22]. PML-RARα and PLZF-RARα can also heterodimerize with PML and PLZF, thus acting, in principle, as double dominant-negative oncogenic products, interfering with both X and RAR/RXR-RA pathways [10,15,23–25]. However, PML-RARα and PLZF-RARα proteins retain intact RARα DNA- and ligand-binding domains, and have an affinity for the ligand comparable to that of the wild-type RARα [26,27].

For these reasons, the molecular mechanisms by which both X-RARα molecules would be leukemogenic at physiological doses of RA, and would behave differently at pharmacological doses of RA, remained until recently unexplained. As a more fundamental corollary, it was also unclear if APL was caused by the aberrant RA-dependent transactivation of gene expression by

X-RARα proteins because, in this case, APL should always be exacerbated by RA, while, on the contrary and paradoxically, RA is extremely effective in APL cases harboring PML-RARα. Thus, it is obvious that the elucidation of the molecular basis of the differential responses to RA in APL is crucial for the understanding of APL pathogenesis itself and goes beyond simply clarifying the mechanisms underlaying RA resistance in APL.

We and others have recently defined the mechanisms of transcriptional repression by X-RARα fusion protein, which in a unified model explain both the molecular pathogenesis of APL and the differential response to RA in APL [28–30]. Our analysis of transgenic mice in which PML-RARα and PLZF-RARα are specifically expressed in the promyelocytic cellular compartment has revealed that the X-RARα fusion proteins play a critical role in leukemogenesis and in determining responses to RA in APL, as *PLZF-RARα* mice develop RA-resistant leukemia while *PML-RARα* mice develop APL-like leukemias that respond to RA [28]. We have demonstrated that both PML-RARα and PLZF-RARα can act as transcriptional repressors and are able to interact with nuclear receptor transcriptional corepressors such as SMRT and N-CoR [28]. PML-RARα can act as a dominant-negative transcriptional repressor of RARα through a nuclear corepressor association that is less sensitive to RA [28]. PLZF-RARα can form, via its PLZF moiety, corepressor complexes that are insensitive to RA [28]. Histone deacetylase inhibitors (HDACIs) such as trichostatin A (TSA), sodium phenylbutyrate (PB), and suberanilohydroxamic acid (SAHA) [31–33], in combination with RA, can overcome the transcriptional repressive activity of PML-RARα and PLZF-RARα [28, and our unpublished observations]. HDACIs can also overcome the unresponsiveness of PLZF-RARα leukemic cells to RA [28]. These findings unravel a crucial role for transcriptional silencing in APL pathogenesis and resistance to RA in APL, and implicate for the first time nuclear receptor corepressors and HDACs in the pathogenesis of human cancer. These observations also suggest that HDACIs alone or in combination with RA might be utilized for the treatment of this leukemia.

"Transcription Therapy" for Cancer

The characterization of our transgenic models of APL demonstrated that high-dose RA therapy is effective in inducing complete clinical remission in leukemia from hCG-PLZF-RARα transgenic mice [28]. This finding is in keeping with the notion that higher RA concentrations might be required to release the PLZF-RARα/nuclear corepressor complex and overcome its dominant negative action on the transcriptional activity of the normal RAR/RXR complexes. It remains to be seen whether in the human t(11;17) a similar

approach might also be effective, because in these patients further unresponsiveness to RA could be conferred by the coexisting RARα-PLZF fusion
protein.

However, a more attractive and general implication of our findings for APL
treatment is the possibility to exploit a combination therapy between
low doses of HDACIs such as TSA, trapoxin, SAHA, PB [31–33], or newly
synthesized HDACIs in combination with RA. This approach would antagonize X-RARα transcriptional repression and concomitantly induce the RA-
dependent transactivation of RA target genes by RAR/RXR or by X-RARα
proteins themselves. It is worth mentioning that specific HDACIs have already
been used in the treatment of malaria, toxoplasmosis, and β-thalassemia
[34,35]. In view of these ideas, we have initiated therapeutic trials in PML-
RARα and PLZF-RARα leukemic mice at presentation, or at relapse upon RA
treatment, with TSA, PB, or SAHA, with or without RA. Despite their general
effects on transcription, the toxicity of HDACIs administered in vivo to mice
is negligible, and indeed, preliminary results obtained in our mouse models
of APL, as well as in human APL patients with combinations of RA and
HDACIs, are extremely promising [36]. This therapeutic approach, which we
termed transcription therapy, represents the first example by which specifically targeting aberrant transcription renders it possible to antagonize the
activity of oncogenic transcription factors (Fig. 1).

It is conceivable that the same therapeutic approach can be applied to other
forms of cancers whereby HDAC-dependent aberrant transcriptional repression is implicated as the main pathogenic mechanism. Straightforward
examples to this end are other forms of myeloid leukemia, non-Hodgkin's
lymphoma, as well as breast and prostate cancer (Table 1). As an example,
non-Hodgkin's lymphomas (NHL), and in particular the diffuse large cells
lymphomas (DLCL) and follicular lymphomas (FL), which are associated with
structural alterations disrupting the BCL6 gene, might represent ideal candidates for "transcription therapy" [37,38]. BCL6, as PLZF, is a transcriptional
repressor that can repress transcription through histone deacetylation by
recruiting nuclear receptor corepressors [39]. The position of the breakpoints

Transcription Therapy

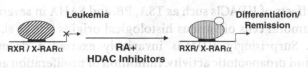

FIG. 1. Transcription therapy with histone deacetylase (HDAC) inhibitors aims at revert-
ing the transcriptional properties of an aberrant oncogenic transcription factor

TABLE 1. Direct applications of "transcription therapy"

Hematological malignancy	Genetic lesion	Mechanism of transcriptional repression
APL	X-RARα	HDAC involved
AML-M2	AML1-ETO	HDAC involved
NHL	BCL-6 rearrangement	HDAC involved

APL, acute promyclocytic leukemia; RAR, retinoic acid receptor; AML, acute myclocytic leukemia; NHL, non-Hodg kin's lymphoma; HDAC, histone deacetylase.

within the BCL6 locus have been mapped: all the observed breakpoints were located within the 5'-flanking region, within the first exon, or within the first intron of BCL6. As a result, the coding region of BCL6 is left intact, whereas the 5'-regulatory region of the gene that contains the promoter sequences is truncated or completely removed.

Analysis of the resulting cDNA predicts that the functional consequence of the translocation is the expression of a normal BCL6 protein under the control of a heterologous promoter leading to the loss of its normal pattern of regulation. This model is valid for various translocations involving 3q27 [40]. As a result, BCL-6 is inappropriately overexpressed within the lymphoid compartment, resulting in aberrant transcriptional repression and lymphoid oncogenic transformation. Thus, it is logical to hypothesize that this form of tumor also might be treated via a therapeutic approach which would block the ability of BCL-6 to repress transcription. An additional straightforward example is represented by acute myeloid leukemias of the M2 subtype [according to the French-American-British (FAB)], associated with the t(8;21) chromosomal translocation involving the *AML1* and *ETO* genes (Table 1). The AML1-ETO fusion protein, unlike the AML-1 protein which is a transcriptional activator, is a potent dominant transcriptional repressor [41,42]. Once again the repressive ability of the AML1-ETO chimeric protein is conferred by its ability to interact, as for the X-RARα proteins of APL, with nuclear corepressors and in turn with HDAC [41,42]. Experiments with HDACIs in cell lines and xenograf mouse models of DLCL and FL and AML-M2 are presently ongoing.

Furthermore, to explore the possibility that HDACIs alone or in combination with RA could be used as anticancer agents, we investigated the biological effects of HDACIs such as TSA, PB, and SAHA in several cancer cell lines and tumoral cells of various histological origins (He et al., manuscript submitted). Surprisingly, HDACIs invariably exerted a dramatic growth inhibitory and proapoptotic activity. Inhibition of proliferation and induction of apoptosis by HDACIs were potentiated by RA. More strikingly, these HDACIs increased RA-induced differentiation. HDACIs, but not RA,

FIG. 2. A model for HDACI tumor suppressive activity: HDACIs specifically derepress a preprogrammed set of genes whose transcriptional activation induces cell cycle arrest, apoptosis, and cellular differentiation

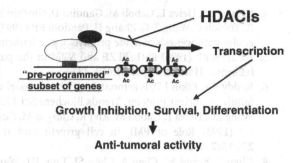

induced detectable hyperacetylation of histones. These findings suggest that HDACIs alone or in combination with RA might be utilized as broad anticancer agents. Based on these findings, we propose a model by which HDACIs specifically derepress a preprogrammed set of genes whose transcriptional activation induces cell cycle arrest, apoptosis, and cellular differentiation (Fig. 2).

Acknowledgments. We thank all the present and past members of the Laboratory of Molecular and Developmental Biology (MADB Lab), at Memorial Sloan-Kettering Cancer Center, working on APL-related subjects: Maria Barna, Mantu Bhaumik, Cristina Cenciarelli, Patricia Clark, Laurent Delva, Mirella Gaboli, Domenica Gandini, Marco Giorgio, Nicola Hawe, Sundeep Kalantry, Taha Merghoub, Daniela Peruzzi, Roberta Rivi, Simona Ronchetti, Eduardo Rego, Davide Ruggero, Carla Tribioli, Zhu-Gang Wang, Hui Zhang, and Sue Zhong, as well as Letizia Longo in our department; in particular, special thanks to Li-Zhen He who almost single-handedly developed what is discussed in this article. We are also grateful to our collaborators on this specific topic: Ray P. Warrell, Jr., Victoria Richon, Richard Rifkind, and Paul Marks. The work carried out in our laboratories is funded by NIH/NCI grants and by the Sloan-Kettering Institute. P.P.P. is a Scholar of the Leukemia Society of America.

References

1. Look AT (1997) Oncogenic transcription factors in the human acute leukemias. Science 278:1059–1064
2. Grignani F, Fagioli M, Alcalay M, Longo L, Pandolfi PP, Donti E, Biondi, A, Lo Coco F, Pelicci PG (1994) Acute promyelocytic leukemia: from genetics to treatment. Blood 83:10–25
3. Warrell RP Jr, de The H, Wang ZY, Degos L (1993) Acute promyelocytic leukemia. New Engl J Med 329:177–189

4. Kalantry S, Delva L, Gaboli M, Gandini D, Giorgio M, Hawe N, He L-Z, Peruzzi D, Rivi R, Triboli C, Wang Z-G, Zhang H, Pandolfi PP (1997) Gene rearrangements in the molecular pathogenesis of acute promyelocytic leukemia. J Cell Physiol 173:288–296

5. Pandolfi PP (1996) PML, PLZF and NPM in the pathogenesis of acute promyelocytic leukemia. Haematologica 81:472–482

6. Reddy BA, Etkin LD, Freemont PS (1992) A novel zinc finger coiled-coil domain in a family of nuclear proteins. Trends Biochem Sci 17:344–345

7. Wang ZG, Delva L, Gaboli M, Rivi R, Giorgio M, Cordon-Cardo C, Grosveld F, Pandolfi PP (1998) Role of PML in cell growth and the retinoic acid pathway. Science 279:1547–1551

8. Chen Z, Brand NJ, Chen A, Chen SJ, Tong JH, Wang ZY, Waxman S, Zelent A (1993) Fusion between a novel Kruppel-like zinc finger gene and the retinoic acid receptor-alpha locus due to a variant t(11;17) translocation associated with acute promyelocytic luekaemia. EMBO J 12:1161–1167

9. Li JY, English MA, Ball HJ, Yeyati PL, Waxman S, Licht JD (1997) Sequence-specific DNA binding and transcriptional regulation by the promyelocytic leukemia zinc finger protein. J Biol Chem 272:22447–22455

10. Koken MHM, Reid A, Quignon F, Chelbi-Alix MK, Davies JM, Kabarowski JHS, Zhu J, Dong S, Chen S, Chen Z, Tan CC, Licht J, Waxman S, de The H, Zelent A (1997) Leukemia-associated retinoic acid receptor alpha fusion partners, PML and PLZF, heterodimerize and colocalize to nuclear bodies. Proc Natl Acad Sci USA 94:10255–10260

11. Chambon P (1996) A decade of molecular biology of retinoic acid receptors. FASEB J 10:940–954

12. Grunstein M (1997) Histone acetylation in chromatin structure and transcription. Nature (Lond) 389:349–352

13. Gudas L, Sporn M, Roberts A (1994) The retinoids: Cellular biology and biochemistry of the retinoids. Raven Press, New York, pp 443–520

14. Licht JD, Chomienne C, Goy A, Chen A, Scott A, Head DR, Michaux JL, Wu Y, DeBlasio A, Miller WH Jr, Zelenetz AD, Willman CL, Chen Z, Chen S-J, Zelent A, Macintyre E, Veil A, Cortes J, Kantarjian H, Waxman S (1995) Clinical and molecular characterization of a rare syndrome of acute promyelocytic leukemia associated with translocation (11;17). Blood 85:1083–1094

15. Perez A, Kastner P, Sethi S, Lutz Y, Reibel C, Chambon P (1993) PML/RAR homodimers: distinct DNA. Binding properties and heteromeric interactions with RXR. EMBO J 12:3171–3182

16. Dong S, Zhu J, Reid A, Strutt P, Guidez F, Zhong H-J, Wang Z-Y, Licht J, Waxman S, Chomienne C, Chen Z, Zelent A, Chen S-J (1996) Amino-terminal protein-protein interaction motif (POZ-domain) is responsible for activities of the promyelocytic leukemia zinc finger-retinoic acid receptor-α fusion protein. Proc Natl Acad Sci USA 93:3624–3629

17. de The H, Lavau C, Marehio A, Chomienne C, Degos L, Dejean A (1991) The PML/RARα fusion mRNA generated by the t(15;17) translocation in acute promyelocytic leukaemia encodes a functionally altered RAR. Cell 66:675–684

18. Pandolfi PP, Grignani F, Alcalay M, Mencarelli A, Biondi A, Lo Coco F, Pelicci PG (1991) Structure and origin of the acute promyelocytic leukemia myl/RARα cDNA and characterization of its retinoid-binding and transactivation properties. Oncogene 6:1285–1292

19. Kastner P, Perez A, Lutz Y, Rochette-Egly C, Gaub M-P, Durand B, Lanotte M, Berger R, Chambon P (1992) Structure, localization and transcriptional properties of two

classes of retinoic acid receptor alpha fusion proteins in acute promyelocytic leukemia (APL): structural similarities with a new family of oncoproteins. EMBO J 11:629–642

20. Kakizuka A, Miller WH Jr, Umesono K, Warrell RP Jr, Frankel SR, Murty VVVS, Dmitrovsky E, Evans RM (1991) Chromosomal translocation t(15;17) in human acute promyelocytic leukemia fuses RARα with a novel putative transcription factor, PML. Cell 66:663–674

21. Chen Z, Guidez F, Rousselot P, Agadir A, Chen SJ, Wang ZY, Degos L, Zelent A, Waxman S, Chomienne C (1994) PLZF-RAR alpha fusion proteins generated from the variant t(11;17)(q23;q21) translocation in acute promyelocytic leukemia inhibit ligand-dependent transactivation of wild-type retinoic acid receptors. Proc Natl Acad Sci USA 91:1178–1182

22. Licht JD, Shaknovich R, English MA, Melnick A, Li J-Y, Reddy JC, Dong S, Chen S-J, Zelent A, Waxman S (1996) Reduced and altered DNA-binding and transcriptional properties of the PLZF-retinoic acid receptor-α chimera generated in t(11;17)-associated acute promyelocytic leukemia. Oncogene 12:323–336

23. Koken MHM, Puvion-Dutilleul F, Guillemin MC, Viron A, Linares-Cruz G, Stuurman N, de Jong L, Szostecki C, Calvo F, Chomienne C, Degos L, Puvion E, de The H (1994) The t(15;17) translocation alters a nuclear body in a retinoic acid-reversible fashion. EMBO J 13:1073–1083

24. Dyck J, Maul GG, Miller WH Jr, Chen JD, Kakizuka A, Evans RM (1994) A novel macro-molecular structure is a target of the promyelocyte-retinoic acid receptor oncoprotein. Cell 76:333–343

25. Weis K, Rambaud S, Lavau C, Jansen J, Carvalho T, Carmo-Fonseca M, Lamond A, Dejean A (1994) Retinoic acid regulates aberrant nuclear localization of PML-RARα in acute promyelocytic luekemic cells. Cell 76:345–356

26. Nervi C, Poindexter EC, Grignani F, Pandolfi PP, Lo Coco F, Avvisati G, Pelicci PG, Jetten AM (1992) Characterization of the PML/RARα chimeric product of the acute promyelocytic leukemia specific t(15;17) translocation. Cancer Res 52:3687–3692

27. Ruthardt M, Testa U, Nervi C, Ferrucci PF, Grignani F, Puccetti E, Grignani F, Peschle C, Pelicci PG (1997) Opposite effects of the acute promyelocytic leukemia PML-retinoic acid receptor alpha (RAR alpha) and PLZF-RAR alpha fusion proteins on retinoic acid signalling. Mol Cell Biol 17:4859–4869

28. He LZ, Guidez F, Tribioli C, Peruzzi D, Ruthardt M, Zelent A, Pandolfi PP (1998) Distinct interactions of PML-RARalpha and PLZF-RARalpha with co-repressors determine differential responses to RA in APL. Nat Genet 18:126–135

29. Grignani F, De Matteis S, Nervi C, Tomassoni L, Gelmetti V, Cioce M, Fanelli M, Ruthardt M, Ferrara FF, Zamir I, Seiser C, Grignani F, Lazar MA, Minucci S, Pelicci PG (1998) Nature 391:815–818

30. Lin RJ, Nagy L, Inoue S, Shao W, Miller WHJ, Evans RM (1998) Nature (Lond) 391:811–814

31. Yoshida M, Horinouchi S, Beppu T (1995) Trichostatin A and trapoxin: novel chemical probes for the role of histone acetylation in chromatin structure and function. Bioessays 17:423–430

32. Lea MA, Tulsyan N (1995) Discordant effects of butyrate analogues on erythroleukemia cell proliferation, differentiation and histone deacetylase. Anticancer Res 15:879–883

33. Richon VM, Emiliani S, Verdin E, Webb Y, Breslow R, Rifkind RA, Marks PA (1998) A class of hybrid polar inducers of transformed cell differentiation inhibits histone deacetylases. Proc Natl Acad Sci USA 95:3003–3007

34. Collins AF, Pearson HA, Giardina P, McDonagh KT, Brusilow SW, Dover GJ (1995) Oral sodium phenylbutyrate therapy in homozygous beta thalassemia: a clinical trial. Blood 85:43–49
35. Darkin-Rattray SJ, Gurnett AM, Myers RW, Dulski PM, Crumley TM, Allocco JJ, Cannova C, Meinke PT, Colletti SL, Bednarek MA, Singh SB, Goetz MA, Dombrowski AW, Polishook JD, Schmatz DM (1996) Apicidin: a novel antiprotozoal agent that inhibits parasite histone deacetylase. Proc Natl Acad Sci USA 93:13143–13147
36. Warrell RP Jr, He L-Z, Richon V, Calleja E, Pandolfi PP (1998) Therapeutic targeting of transcription in acute promyelocytic leukemia by use of an inhibitor of histone deacetylase. J Natl Cancer Inst 90:1621–1625
37. Lo Coco F, Ye BH, Lista F, Corradini P, Offit K, Knowles DM, Chaganti RSK, Dalla-Favera R (1994) Rearrangements of the BCL-6 gene in diffuse large-cell non-Hodgkins lymphoma. Blood 83:1757–1759
38. Magrath I (1990) Lymphocyte ontogeny: a conceptual basis for understanding neoplasia of the immune system. In: Magrath I (ed) The non-Hodgkin's Lymphoma. Williams and Wilkins, Baltimore, pp 29–48
39. Dhordain P, Albagli O, Lin RJ, Ansieau S, Quief S, Leutz A, Kerckaert JP, Evans RM, Leprince D (1997) Corepressor SMRT binds the BTB/POZ repressing domain of the LAZ3/BCL6 oncoprotein. Proc Natl Acad Sci USA 94:10762–10767
40. Ye BH, Chaganti S, Chang C-C, Niu H, Corradini P, Chaganti RSK, Dalla Favera R (1995) Chromosomal translocations cause deregulated BCL-6 expression by promoter substitution in B-cell lymphoma. EMBO J 14:6209–6217
41. Wang J, Hoshino T, Redner RL, Kajigaya S, Liu JM (1998) ETO, fusion partner in t(8;21) acute myeloid leukemia, represses transcription by interaction with the human N-CoR/mSin3/HDAC1 complex. Proc Natl Acad Sci USA 95:10860–10865
42. Gelmetti V, Zhang J, Fanelli M, Minucci S, Pelicci PG, Lazar MA (1998) Aberrant recruitment of the nuclear receptor corepressor–histone deacetylase complex by the acute myeloid leukemia fusion partner ETO. Mol Cell Biol 18:7185–7191

A Novel Acute Promyelocytic Leukemia Model: From Cell to Murine System

Masahiro Kizaki

Summary. Differentiation-inducing therapy by all-*trans* retinoic acid (RA) is now a standard therapy in patients with acute promyelocytic leukemia (APL). Nearly all patients achieve complete remission by the treatment of all-*trans* RA; however, clinical remissions are usually of brief duration, and these patients often develop RA-resistant disease. The mechanisms of RA resistance in APL cells are poorly understood, and most clinical approaches have not been successful in overcoming RA resistance. We have recently established a novel APL cell line (UF-1) with RA-resistant features. In addition, we have established a human GM-CSF-producing transgenic (hGMTg) SCID mice system. UF-1 cells were inoculated either intraperitoneally or subcutaneously into hGMTg SCID mice and to make the first RA-resistant murine APL model. These RA-resistant APL model systems in vitro and in vivo may be useful for investigating molecular studies on blocking of leukemic cell differentiation and as a means to investigate the mechanisms of RA resistance. Moreover, this murine model system will be important for developing novel therapeutic strategies in RA-resistant APL.

Key words. Retinoic acid, APL, Differentiation, RA resistance, Murine model

Retinoic Acid Resistance in Acute Promyelocytic Leukemia

Acute promyelocytic leukemia (APL) is associated with a specific 15;17 translocation that generates a PML/RARα fusion protein between the PML gene on chromosome 15 and the RARα gene on chromosome 17 [1,2].

Division of Hematology, Keio University School of Medicine, 35 Shinanomachi, Shinjuku-ku, Tokyo 160-8582, Japan

PML/RARα retains a variety of functionally important domains of both PML and RARα proteins, such as three cysteine/histidine-rich regions including a RING-finger domain and ligand-binding domain, respectively. The PML/RARα fusion protein has been proposed to block myeloid differentiation via inhibition of nuclear receptor response, and subsequently to be involved in the leukemogenesis of APL [3,4]. Several clinical studies have shown that nearly all patients who are treated with all-*trans* retinoic acid (RA) achieve complete remission through an induction of differentiation of the leukemic cells [5–8]. However, clinical remissions in patients treated with all-*trans* RA alone are usually of brief duration, and these patients often develop RA-resistant disease [4]. Approximately 20–30% of patients treated with first-line approaches have been reported to relapse, and relapsed APL patients more frequently show refractory disease or a short-lived second remission [9–11]. Therefore, clinical resistance to RA poses a serious problem for differentiation-inducing therapy. Several hypotheses are proposed to explain the development of RA resistance in APL patients [12]. One potential mechanism for resistance to RA is pharmacological alterations in the metabolism of all-*trans* RA by induction of cytochrome P450 and cellular RA-binding protein II (CRABPII) or enhanced RA excretion by induction of P-glycoprotein, resulting in lowering plasma and intracellular levels of active retinoids [13,14]. The other obvious explanation is the generation of new mutations in the retinoid receptors. Recent studies revealed that several clinical samples from RA-resistant APL patients exhibit point mutations in the ligand-binding domain of the PML/RARα gene [15,16]. These results suggest that the point mutation of the chimeric gene is important for the development of acquired RA resistance, although its clinical significance is not yet clear.

RA-Resistant APL Cell Line, UF-1

To understand the characteristics of leukemic cells, human cell lines are useful tools for study. The APL cells have strikingly low proliferation potential in vitro; thus, only one APL cell line has been isolated, designated NB4, which can be differentiated to granulocytes by all-*trans* RA [17]. In contrast, we have developed a novel APL cell line (UF-1) from a patient who was clinically resistant to all-*trans* RA; therefore, this cell line reflects RA-resistant clinical situations [18]. UF-1 cells showed considerable heterogeneity in cell size and nuclear cytoplasmic ratio (N/C). The nucleoli were large, round, and often lobulated, and a few nucleoli were found per cell (Fig. 1). Numerous granules were present in all the cells. NB4 cells were treated with 10^{-8} M all-*trans* RA for 4 days resulted in differentiation toward mature granulocytes (Fig. 1). In contrast, all-*trans* RA did not induce morphological differentiation of UF-1 cells

Control   **All-*trans* RA**

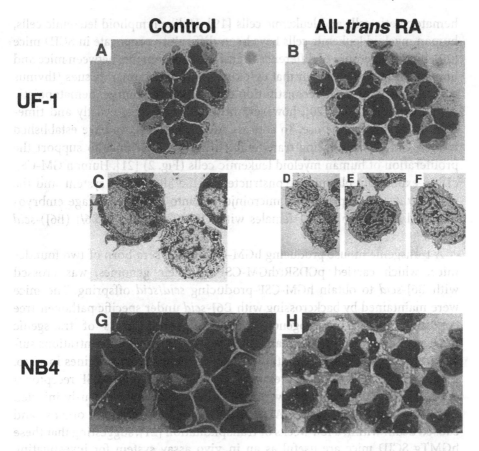

UF-1

NB4

FIG. 1A–H. Morphological changes in UF-1 and NB4 cells by all-*trans* retinoic acid (RA) (10^{-8} M). Cytospine slides were prepared and stained with Wright-Giemsa (UF-1, A, B; NB4, G, H; ×1 000). Electron photo micro graphs of UF-1 cells (C, ×5 000; D, E, F, ×3 300)

toward either mature granulocytes or monocytes (Fig. 1). These results suggest that the UF-1 cell is the first permanent APL cell line with spontaneous RA resistance. Therefore, this RA-resistant APL cell line may be a useful model for molecular studies on the blockage of leukemic cell differentiation, as well as a means to investigate the mechanisms of retinoid resistance in leukemic cells.

Human GM-CSF-Producing Transgenic SCID Mice

The transplantation of human hematopoietic cells and fetal organs into severe combined immunodeficiency (SCID) mice has enabled the development of new in vivo models, called SCID-hu mice, for the growth of normal human hematopoietic cells and leukemic cells [19]. Unlike lymphoid leukemic cells, human myeloid leukemic cells have been difficult to propagate in SCID mice [20], possibly because of differences of microenvironments between mice and humans. It has been shown that cytokines and fetal human tissues (thymus and bone) increase the reconstitution capacity of the human hematopoietic system in SCID mice [20]; however, these approaches are costly and time-consuming for routine use. To address these problems, we have established human GM-CSF-producing transgenic (hGMTg) SCID mice to support the proliferation of human myeloid leukemic cells (Fig. 2) [21]. Human GM-CSF cDNA expressing plasmid is constructed in the SRα vector system, and the pCDSRαhGM-CSF plasmid is microinjected into prenuclear stage embryos obtained by crossing BDF1 females with C.B-17-scid or C57B6/J (B6J)-*scid* males.

A transgenic mouse producing hGM-CSF in the sera born of two founder mice, which carried pCDSRαhGM-CSF in their genomes, was crossed with B6J-*scid* to obtain hGM-CSF-producing *scid/scid* offspring. The mice were maintained by backcrossing with B6J-*scid* under specific pathogen-free conditions (Fig. 3). Endogenous hGM-CSF serum levels of transgenic mice were measured by ELISA and often reached 1 ng/ml, concentrations sufficient to support the proliferation of GM-CSF-dependent cell lines in vitro. Murine Ba/F3 cells stably expressing functional human GM-CSF receptor α and β chains (BAF/αβ cells) were established and subcutaneously injected into hGMTg SCID mice. BAF/αβ cells invaded multiple organs and caused death within a few weeks of transplantation [21], suggesting that these hGMTg SCID mice are useful as an in vivo assay system for investigating leukemogenesis.

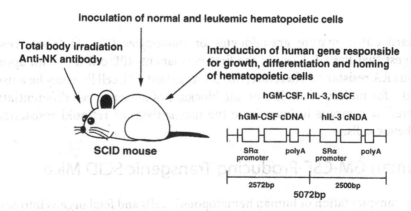

FIG. 2. SCID-hu mouse model with human hematopoietic function. Human GM-CSF cDNA and human IL-3 cDNA are independently driven by SRα promoters

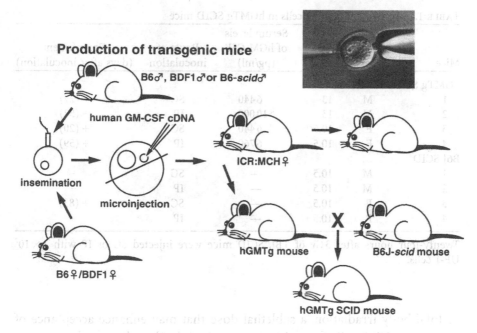

FIG. 3. Establishment of human GM-CSF-producing transgenic SCID mice. Details are shown in Ref. 21

RA-Resistant APL Model in hGMTg SCID Mice

The transgenic models of acute leukemias will be invaluable in the study of the pathogenesis and new therapeutic approaches for the disease. Several transgenic mice have been developed for the PML/RARα chimeric gene, the molecular basis for APL, but some important differences exist between the phenotypes of these early-model mice and humans [22,23]. To resolve these problems, several groups have generated a cathepsin-G (CG) minigene expression vector and a human migration inhibitory factor-related protein (hMRP8) expression cassette, which together drive promyelocyte-specific expression of the targeted gene in mice [24–26]. Of the mice produced from PML/RARα transgenic mice using either hCG or hMRP8 expression vector, 10–20% developed, with a long latency period, to an acute myeloid leukemia characterized by features of APL and responded to retinoids. These transgenic approaches, however, take a long time to produce an APL model and yield only RA-sensitive APL mice. To make an RA-resistant APL model, therefore, we inoculated UF-1 cells into hGMTg SCID mice. To date, there has been one APL mouse model using RA-sensitive NB4 cells and SCID mice [27]. UF-1 cells (1×10^7 cells) were inoculated either intraperitoneally or subcutaneously into hGMTg SCID and control B6J SCID mice. Mice were pretreated with 3 Gy

TABLE 1. Engraftment of UF-1 cells in hGMTg SCID mice

Mice	Sex	Age (weeks)	Serum levels of hGM-CSF (pg/ml)	Route of inoculation	Engraftment (days after inoculation)
hGMTg SCID					
1	M	13	6440	SC	+ (27)
2	M	13	>10000	IP	+ (52)
3	F	10.5	6440	SC	+ (20)
4	F	10.5	6760	IP	+ (59)
B6J SCID					
1	M	10.5	—	SC	−
2	M	10.5	—	IP	−
3	F	10.5	—	SC	+ (83)
4	F	10.5	—	IP	−

Twenty-four hours after 3 Gy of TBI, SCID mice were injected SC or IP with 1×10^7 UF-1 cells.

of total-body irradiation, a sublethal dose that may enhance acceptance of xenografts. UF-1 cells formed tumors at the injection site as subcutaneous tumors in all hGMTg SCID mice, but not control SCID mice, between days 20 and 59 (Table 1). Single-cell suspensions from tumors were similar in morphological, immunological, cytogenetic, and molecular genetic features to parental UF-1 cells. All-*trans* RA did not respond to the suspension cells, indicating that these cells are resistant to RA. Thus, these mice are the first human RA-resistant APL animal model.

Arsenic Trioxide-Induced Apoptosis and Differentiation in the RA-Resistant APL Model

Recent clinical studies in China and the United States have shown that arsenic trioxide (As_2O_3) is an effective treatment of APL patients refractory to all-*trans* RA [28,29]. As_2O_3 has been shown to induce apoptosis of NB4 cells, in association with the downregulation of *bcl-2* gene expression and modulation of PML/RARα chimeric protein [30]. Moreover, it has been demonstrated that As_2O_3 can trigger both apoptosis at higher concentrations (0.5–2 μM) and cellular differentiation at lower concentrations (0.1–0.5 μM) in RA-susceptible NB4 cells and RA-resistant NB4 sublines [30]. However, there have been no studies on As_2O_3 using primary RA-resistant APL cells from patients in vitro or an animal model in vivo. Moreover, arsenicals are considered to be a poison and an environmental carcinogen for human organs. Also, As_2O_3 possibly induces serious side effects in neurogenic toxicity, gastrointestinal and liver dysfunctions. Therefore, it is important to address the effects of As_2O_3 on clinically acquired RA-resistant APL in vitro and in vivo. As_2O_3 can induce inhibition of cellular growth of both RA-sensitive NB4 and RA-resistant UF-1 cells via induction of apoptosis in vitro [31].

The expression of BCL-2 protein decreased in a dose- and time-dependent manner in NB4 cells. Interestingly, the levels of BCL-2 were not modulated by As_2O_3 but it did upregulate BAX protein in UF-1 cells. After 40 days of implantation of UF-1 cells into hGMTg SCID mice, mice were treated with As_2O_3, all-*trans* RA, and PBS for 21 days. In all-*trans* RA- and PBS-treated mice, tumors grew rapidly, with a 4.5-fold increase in volume at day 21 compared to the initial size. In marked contrast, tumor size was decreased to half the initial size by treatment with As_2O_3, which resulted in cells with the typical appearance of apoptosis [31]. Interestingly, one of the As_2O_3-treated mice showed mature granulocytes in the diminished tumor, suggesting that arsenicals had dual effects on RA-resistant APL in vivo, inducing both apoptosis and differentiation of the leukemic cells. Because hGMTg SCID mice had a high amount of human GM-CSF in their sera, GM-CSF may have some effects on As_2O_3-induced apoptosis in RA-resistant APL cells. In vitro experiments revealed that the combination of GM-CSF and As_2O_3 induced differentiation but not apoptosis of UF-1 cells by modulating Jak 2 kinase and MCL-1 protein [32]. Therefore, the combination of heavy metal and cytokine such as As_2O_3 and GM-CSF will provide a novel approach for differentiation-inducing therapy in patients with acute leukemia. Hopefully, this approach will enhance positive responses and will also limit the toxicity of As_2O_3.

In conclusion, our in vitro and in vivo RA-resistant APL model system will be useful for investigating the molecular biology of this myeloid leukemia, as well as for evaluating novel therapeutic approaches, including As_2O_3 therapy to patients with RA-resistant APL.

Acknowledgments. The author thanks the members of the Division of Hematology, Keio University School of Medicine: Kentaro Kinjo, M.D., Hiromichi Matsushita, M.D., Akihiro Muto, M.D., Yumi Fukuchi, Ph.D., Norihiro Awaya, M.D., Chiharu Kawamura, M.D., Nobuyuki Takayama, M.D., David T. Kamata, M.D., Yutaka Hattori, M.D., Hideo Uchida, M.D., and Yasuo Ikeda, M.D. The author also thanks to Mamoru Ito, Ph.D. (Central Institute for Experimental Animals, Kanagawa, Japan) and Yoshito Ueyama, M.D. (Department of Pathology, Tokai University School of Medicine, Kanagawa, Japan) for providing human GM-CSF-producing transgenic SCID mice. This work was supported by grants from the Ministry of Education, Science and Culture in Japan and the National Grant-in Aid for the Establishment of High-Tech Research Center in a Private University.

References

1. de The H, Chomienne C, Lanotte M, Degos L, Dejean A (1991) The t(15;17) translocation of acute promyelocytic leukemia fuses the retinoic acid receptor α gene to a novel
   transcribed locus. Nature (Lond) 347:558–561
2. Kakizuka A, Miller WH Jr, Umesono K, Warrell RP Jr, Frankel SR, Murty VVVS,
   Dmitrovsky E, Evans RM (1991) Chromosomal translocation t(15;17) in human acute
   promyelocytic leukemia fuses RARα with a novel putative transcription factor, PML.
   Cell 66:663–674
3. Grignani F, Ferrucci PF, Testa U, Talamo G, Fagioli M, Alcalay M, Mencarelli A,
   Grignani F, Peschle C, Nicoletti I, Pelicci PG (1993) The acute promyelocytic leukemia
   specific PML/RARα protein inhibits differentiation and promotes survival of myeloid
   precursor cells. Cell 74:423–431
4. Warrell RP Jr, de The H, Wang ZY, Degos L (1993) Acute promyelocytic leukemia.
   N Engl J Med 329:177–189
5. Huang ME, Ye YC, Chen SR, Chai JR, Lu JX, Zhoa L, Gu HT, Wang ZY (1988) Use of all-
   trans retinoic acid in the treatment of acute promyelocytic leukemia. Blood 72:567–572
6. Castaigne S, Chomienne C, Daniel MT, Ballarini P, Berger R, Fenaux P, Degos L (1990)
   All-*trans* retinoic acid as a differentiation therapy for acute promyelocytic leukemia.
   I. Clinical results. Blood 76:1704–1709
7. Warrell RP Jr, Frankel SR, Miller WH Jr, Sheinberg DA, Itri LM, Hittelman WN, Vyas
   R, Andreeff M, Tafuri A, Jakubowski A, Gabrilove J, Gordon MS, Dmitrovski E (1991)
   Differentiation therapy of acute promyelocytic leukemia with tretinoin (all-*trans*-
   retinoic acid). N Engl J Med 324:1385–1393
8. Kanamaru A, Takemoto Y, Tanimoto M, Murakami H, Asou N, Kobayashi T, Kuriyama
   K, Ohmoto E, Sakamaki H, Tsubaki K, Hiraoka H, Yamada O, Oh H, Saito K, Matsuda
   S, Minato K, Ueda T, Ohno R (1995) All-*trans* retinoic acid for the treatment of newly
   diagnosed acute promyelocytic leukemia. Blood 85:1202–1206
9. Fenaux P, Chomienne C, Degos L (1997) Acute promyelocytic leukemia: biology and
   treatment. Semin Oncol 24:92–102
10. Mandelli F, Diverio D, Avvisati G, Luciano A, Barbui T, Bernasconi C, Broccia G,
    Cerri R, Falda M, Fioritoni G, Leoni F, Liso V, Petti MC, Rodeghiero F, Saglio G,
    Vegna ML, Visani G, Jehn U, Willemze R, Muus P, Pelicci PG, Biondi A, Lo Coco F
    (1997) Molecular remission in PML/RARα-positive acute promyelocytic leukemia
    by combined all-*trans* retinoic acid and idarubicin (AIDA) therapy. Blood
    90:1014–1021
11. Tallman MS, Andersen JW, Schiffer CA, Appelbaum FR, Feusner JH, Ogden A,
    Shepherd L, Willman C, Bloomfield CD, Rowe JM, Wiernik PH (1997) All-*trans* retinoic
    acid in acute promyelocytic leukemia. N Engl J Med 337:1021–1028
12. Kizaki M, Ueno H, Matsushita H, Takayama N, Muto A, Awaya N, Ikeda Y (1997)
    Retinoid resistance in leukemic cells. Leuk Lymphoma 25:427–434
13. Kizaki M, Ueno H, Yamazoe Y, Shimada M, Takayama N, Muto A, Matsushita H,
    Nakajima H, Morikawa M, Koeffler HP, Ikeda Y (1996) Mechanisms of retinoid resis-
    tance in leukemic cells: possible role of cytochrome P450 and P-glycoprotein. Blood
    87:725–733
14. Delva L, Cornic M, Balitrand N, Guidez F, Miclea J, Delmar A, Teillet F, Fenaux P,
    Castaigne S, Degos L, Chomienne C (1993) Resistance to all-*trans* retinoic acid (ATRA)
    therapy in relapsing acute promyelocytic leukemia: study of in vitro sensitivity and
    cellular binding protein levels in leukemic cells. Blood 82:2175–2181

15. Imaizumi M, Suzuki H, Yoshinari M, Sato A, Saito T, Sugawara A, Tsuchiya S, Hatae Y, Fujimoto T, Kakizuka A, Konno T, Iinuma K (1998) Mutations in the E-domain of RARα portion of the PML/RARα chimeric gene may confer clinical resistance to all-*trans* retinoic acid in acute promyelocytic leukemia. Blood 92:374–382

16. Ding W, Li YP, Nobile LH, Grills G, Carrera I, Paietta E, Tallman MS, Wiernik PH, Gallagher RE (1998) Leukemic cellular retinoic acid resistance and missense mutations in the PML-RARα fusion gene after relapse of acute promyelocytic leukemia from treatment with all-trans retinoic acid and sensitive chemotherapy. Blood 92:1172–1183

17. Lanotte M, Martin-Thouvenin V, Najman S, Balerini P, Valensi F, Berger R (1991) NB4, a maturation inducible cell line with t(15;17) marker isolated from a human acute promyelocytic leukemia (M3). Blood 77:1080–1086

18. Kizaki M, Matsushita H, Takayama N, Muto A, Ueno H, Awaya N, Kawai Y, Asou H, Kamada N, Ikeda Y (1996) Establishment of a novel acute promyelocytic leukemia cell line (UF-1) with retinoic acid-resistant features. Blood 88:1824–1833

19. Dick JE, Pflumio F, Lapidot T (1991) Mouse models for human hematopoiesis. Semin Immunol 3:367–378

20. Uckun FM (1996) Severe combined immunodeficient mouse models of human leukemia. Blood 88:1135–1146

21. Miyakawa Y, Fukuchi Y, Ito M, Kobayashi K, Kuramochi K, Ikeda Y, Takebe Y, Tanaka T, Miyasaka M, Nakahata T, Tamaoki N, Nomura T, Ueyama Y, Shimamura K (1996) Establishment of human granulocyte-macrophage colony stimulating factor producing transgenic SCID mice. Br J Haematol 95:437–442

22. Altabef M, Garcia M, Lavau C, Bae S-C, Dejean A, Samarut J (1996) A retrovirus carrying the promyelocyte-retinoic acid receptor PML-RARα fusion genes transforms hematopoietic progenitors in vitro and induces acute leukemia. EMBO J 15:2707–2716

23. Early E, Moore MAS, Kakizuka A, Nason-Burchenal K, Martin P, Evans RM (1996) Transgenic expression of PML-RARα impairs myelopoiesis. Proc Natl Acad Sci USA 93:7900–7904

24. He L-Z, Tribioli C, Rivi R, Peruzzi D, Pelicci PG, Soares V, Cattoretti C, Pandolfi PP (1997) Acute leukemia with promyelocytic features in PML/RARα transgenic mice. Proc Natl Acad Sci USA 94:5302–5307

25. Grisolano JL, Wesselschmidt RL, Pelicci PG, Ley TJ (1997) Altered myeloid development and acute leukemia in transgenic mice expressing PML-RARα under control of cathepsin G regulatory sequences. Blood 89:376–387

26. Brown D, Kogan S, Lagasse E, Weissman I, Alcalay M, Pelicci PG, Atwater S, Bishop JM (1997) A PMLRARα transgene initiate murine promyelocytic leukemia. Proc Natl Acad Sci USA 94:2551–2556

27. Zhang SY, Zhu J, Chen GQ, Du XX, Lu LJ, Zhang Z, Zhang HJ, Chen HR, Wang ZY, Berger R, Lanotte M, Waxman S, Chen Z, Chen SJ (1996) Establishment of a human promyelocytic leukemia-ascites model in SCID mice. Blood 87:3404–3409

28. Shen ZX, Chen GQ, Ni JH, Li XS, Xiong SM, Qiu QY, Zhu J, Tang W, Sun GL, Yang KQ, Chen Y, Zhou L, Fang ZW, Wang YT, Ma J, Zhang P, Zhang TD, Chen SJ, Chen Z, Wang ZY (1997) Use of arsenic trioxide (As$_2$O$_3$) in the treatment of acute promyelocytic leukemia (APL): II. Clinical efficiency and pharmacokinetics in relapsed patients. Blood 89:3354–3360

29. Soignet SL, Maslak P, Wang ZG, Thanwar S, Calleja E, Dardashti LJ, Corso D, DeBlasio A, Gabrilove J, Scheinberg DA, Pandolfi PP, Warrell RP Jr (1998) Complete remission after treatment of acute promyelocytic leukemia with arsenic trioxide. N Engl J Med 339:1341–1348

30. Chen GQ, Zhu J, Shi XG, Ni JH, Zhong HJ, Si GY, Jin XL, Tang W, Li XS, Xong SM, Shen ZX, Sun GL, Ma J, Zhang P, Zhang TD, Gazin C, Naoe T, Chen SJ, Wang ZY, Chen Z (1996) In vitro studies on cellular and molecular mechanisms of arsenic trioxide (As₂O₃) in the treatment of acute promyelocytic leukemia: As₂O₃ induces NB4 cells apoptosis with downregulation of bcl-2 expression and modulation of PML-RARα/PML proteins. Blood 88:1052–1061

31. Kinjo K, Kizaki M, Muto A, Fukuchi Y, Umezawa A, Yamato K, Nishihara T, Hata J, Ito M, Ueyama Y, Ikeda Y (1999) Arsenic trioxide (As₂O₃)-induced apoptosis and differentiation in retinoic acid-resistant acute promyelocytic leukemia model in hGM-CSF-producing transgenic SCID mice. Leukemia (in press)

32. Kizaki M, Muto A, Kinjo K, Ueno H, Ikeda Y (1998) Application of heavy metal and cytokine for differentiation-inducing therapy in acute promyelocytic leukemia. J Natl Cancer Inst 90:1906–1907

Phase 1 Trial of Retroviral-Mediated Transfer of the Human MDR-1 in Patients Undergoing High-Dose Chemotherapy and Autologous Stem Cell Transplantation

CHARLES HESDORFFER, JANET AYELLO, MAUREEN WARD,
ANDREAS KAUBISCH, ARTHUR BANK, and KAREN ANTMAN

Summary. Normal bone marrow cells have little or no expression of the MDR p-glycoprotein product and are, therefore, particularly susceptible to killing by MDR-sensitive drugs such as the vinca alkaloids, anthracyclines, podophyllotoxins, and taxanes. In this report of a phase 1 clinical study performed at the Columbia-Presbyterian Medical Center, we demonstrate the safety and efficacy of transfer of the human multiple drug resistance (MDR) gene into hematopoietic stem cells and progenitors in bone marrow as a means of providing resistance of these cells to the toxic effects of cancer chemotherapy. One-third of the cells harvested from patients undergoing autologous bone marrow transplantation as part of high-dose chemotherapy treatment for advanced cancer were transduced with an MDR cDNA-containing retrovirus; these transduced cells were reinfused together with unmanipulated cells following the administration of the high-dose chemotherapy. High-level MDR transduction of BFU-E and CFU-GM derived from transduced CD34+ cells was demonstrated posttransduction and prereinfusion. However, only two of the five patients showed evidence of MDR transduction of their marrow at a low level at 10 weeks and 3 weeks, respectively, post transplantation. This relatively unexpected low level of efficiency of transduction was thought to be because the unmanipulated cells, infused at the same time as the transduced cells, might compete with the cytokine-stimulated transduced cells in repopulating the marrow. The MDR retroviral supernatant used was shown to be free of replication-competent retrovirus (RCR) before use, and all tests of patient samples post transplantation were negative for RCR. In addition, no adverse events with respect to marrow engraftment or other problems related

Divisions of Hematology and Oncology, Department of Medicine and Genetics and Development, College of Physicians and Surgeons of Columbia University, 177 Fort Washington Avenue, New York, NY 10032, USA

to marrow transplantation were encountered. This study does indicate the feasibility and safety of bone marrow gene therapy with a potentially therapeutic gene, the MDR gene.

Key words. Gene therapy, MDR, Retrovirus, Advanced malignancies

Introduction

High-dose chemotherapy followed by autologous marrow or peripheral blood progenitor cell transplantation has become an accepted therapy for selected patients with various malignancies who have a low probability of cure with conventional therapy. In the laboratory, a linear increase in the dose of various chemotherapeutic agents causes a log increase in the death rate of cancer cells [1]. The myeloablative effect of increasing doses of chemotherapy can be ameliorated by the reinfusion of hematopoietic progenitor cells harvested before the administration of chemotherapy from either marrow or peripheral blood. In retrospective analyses, the administration of the full planned dose of chemotherapy, in comparison with patients for whom doses were attenuated for various reasons, results in an improved disease-free and overall survival [2,3]. In small randomized trials in myeloma, relapsed non-Hodgkin's lymphoma, and breast cancer, high-dose chemotherapy with hematopoietic stem cell support has increased the disease-free survival rate over conventional dose therapy [4–6].

An increased number of cycles of high-dose chemotherapy might further improve survival rates, but are often limited by repeated pancytopenia with associated life-threatening infection or hemorrhage. To resolve this problem, gene transfer strategies aimed at making hematopoietic stem cells and progenitor cells resistant to specific chemotherapy drugs have been devised and tested in animal models and in human hematopoietic progenitor and stem cells (HPC) in vitro [7–11]. We report here our first clinical trial with retroviral transduction of the human multiple drug resistance (MDR-1, *MDR*) gene into marrow- or blood-derived progenitor cells as a potential means of protecting these cells from the myelotoxic effects of chemotherapy [12]. The MDR gene is normally expressed at low levels in HPC, and thus these cells are particularly susceptible to the myelotoxic effects of many anticancer drugs including the vinca alkaloids, anthracyclines, podophyllins, and paclitaxel and its congeners, all of which require the MDR gene product, *p*-glycoprotein, for their removal from the cells. The aim of this phase I study is to evaluate the feasibility and toxicity associated with the transduction of the MDR gene into HPC for patients with solid tumors undergoing high-dose chemotherapy with hematopoietic stem cell support.

Patients and Methods

Patient Selection

Patients with histologically confirmed and advanced breast, ovarian, and brain tumors were eligible for this study if they had been entered on an institutional high-dose protocol requiring hematopoietic stem cell support. At least a partial response to conventional dose chemotherapy and no lesions on bone scan were required for patients with metastatic breast cancer. For ovarian cancer, patients with stage III or IV and recurrent or residual disease after debulking surgery and at least three cycles of a standard platinum-based chemotherapy were eligible. Brain tumor patients were eligible if they had recurrent anaplastic oligodendrogliomas or anaplastic astrocytomas or glioblastomas at any point in their management. Further eligibility criteria included ages 18 to 65; normal renal, cardiac, hepatic, and pulmonary function; normal marrow cellularity; and no tumor on marrow aspirate and biopsy. Four patients with breast cancer, two with ovarian cancer, and one patient with a glioblastoma were enrolled. This protocol was approved by the Columbia-Presbyterian Medical Center (CPMC), IRB, the FDA, and the NIH committee on Recombinant DNA Research (RAC). All patients signed consent to undergo gene transduction as well as the disease-specific high-dose chemotherapy protocol used to treat their malignant process.

Preparation of MDR Gene-Containing Retroviral Supernatants

The amphotropic MDR viral producer line (A12M1) used in these studies has been previously described [9,11]. These are NIH 3T3 cells with split retroviral genomes in GP+Am12 [13], also containing pHAMDR1, a plasmid with a retroviral vector with an MDR cDNA [7,14]. A12M1 cells were grown in 175-cm^3 flasks (Nunc, Napervelle, IL, USA) with Dulbecco's modified Eagle's media (DME; Gibco BRL), 10% newborn calf serum, and 1% antibiotic/antimycotic (Sigma, St. Louis, MO, USA). At 80% confluence, the DME media was removed, the cells were washed with phosphate-buffered saline (PBS), and replaced with Iscove's modified Dulbecco's media (IMDM; Gibco BRL), 20% fetal calf serum (FCS), and 1% penicillin/streptomycin (Sigma) [9]. After 24h, viral supernatant was collected and filtered through a 0.45-μm filter and frozen at −70°C. Clinical-grade A12M1 supernatants were supplied by Genetix Pharmaceuticals (Cambridge, MA, USA) and prepared by Magenta Rockville, MD, USA).

Hematopoietic Cell Transduction

Transduction of bone marrow was performed in the first six patients. Protocol amendment allowed for blood-derived stem cell transduction in patient 7. Marrow was harvested from the posterior iliac crests at least 2 weeks after any chemotherapy or growth factor treatment. Peripheral blood stem cells were mobilized with chemotherapy and G-CSF and harvested using either a Fenwall CS3000 or COBE Spectra beginning when the peripheral white cell count was >5000/µl. Two-thirds of the hematopoietic stem cells or marrow containing at least 1×10^6 CD34+ cells/kg were collected and frozen without further manipulation. The remaining third of the harvested cells were processed to select CD34+ cells using either the Applied Immune Sciences plate method [15,16] or the CellPro CEPRATE™ stem cell concentrator column [17–19]. The yields of CD34+ cells were much greater using the CellPro columns and the procedure much simpler so that these columns were used in all patients after patient 1. The percent transduced CD34+ cells per total CD34+ cells reinfused into the patients was calculated from the number of cells obtained from peripheral blood and marrow and by the percent CD34+ cells assessed by fluorescence-activated cell sorting (FACS) analysis of transduced samples.

The isolated CD34+ cells were incubated for 48h on fibronectin-coated plates (Collaborative) with clinical-grade 100U/ml IL-3, 200U/ml IL-6, and 50ng/ml human stem cell factor (SCF; Amgen) [9]. Cells were centrifuged for 10min at $8000 \times g$, resuspended in A12M1 supernatants mixed with the same concentrations of growth factors and 4mg/ml of sterile-filtered polybrene, and replated on fibronectin-coated plates for 8–10h. Cells were again collected, centrifuged, and resuspended in a fresh A12M1/growth factor mixture overnight. The transduced cells were then cryopreserved. Between 372 and 2750 ml of A12M1 supernatant with a titer of 2×10^4 was added to $1.6–14 \times 10^7$ CD34+ cells to assure a multiplicity of infection of close to 1 in most patients.

High-dose chemotherapy administration, isolation procedures, and prophylactic antibiotics (ciprofloxacin, fluconazole, and acyclovir for patients with herpes simplex virus antibodies) followed standard practice for autologous marrow transplantation at our institution. High-dose chemotherapy was infused over a 96-h period and stem cells were reinfused 48–72h after completion of the chemotherapy. Irradiated blood and platelet transfusions, CMV-negative for CMV-negative individuals, were infused as necessary with a leukocyte depletion filter.

Evaluation for Exogenous MDR Transduction and Expression

Posttransplantation marrow and blood samples were evaluated with MDR-PCR. A set of primers specific for the transduced MDR cDNA from adjacent coding exons was used to differentiate between the retrovirally inserted MDR cDNA and the endogenous MDR gene [7,20]. DNA for PCR was prepared from lysates using phenol/chloroform and ethanol precipitation. A negative control cell sample was processed at the same time to ensure no cross-contamination of samples and solutions. PCR was carried out on 1µg DNA with 1 unit of

AmpliTaq polymerase and appropriate reaction mix (Perkin Elmer/Cetus) using 20 mM of the following MDR cDNA sequence primers: sense-strand, CCCATCATTGCAATAGCAG (residues 2596–2615), and antisense strand GTTCAAACTTCTGCTCCTGA (residues 2733–2752) [7,20]. A 167-bp product specific for the MDR cDNA was amplified with 35 cycles of 30 s denaturation at 94°C, 30 s of annealing at 55°C, and 60 s of extension at 72°C; 10 ml of each reaction was run on a 4% NuSieve (FMC) agarose gel. Methylcellulose colonies were analyzed by PCR as described with the following exceptions: individual colonies were pipetted from the semisolid culture and washed in 1 ml 1 × PBS [11]. The resulting cell pellet was lysed in 50 µl proteinase K buffer as described and subjected to 35–45 cycles of MDR-PCR; 20 µL of this reaction was analyzed on a gel.

The level of surface expression and function of the MDR p-glycoprotein protein produced by the transduced MDR gene was assessed by FACS analysis (FACScan; Beckman Dickinson, Mountainview, CA, USA) using a monoclonal antibody 17F9, which reacts with an external epitope of the p-glycoprotein [7,21,22], and by methylcellulose clonogenic assays for BFU-E and CFU-GM of transduced cells either without or with exposure to 10^{-8} M paclitaxel [11]. In vitro clonogenic assays (BFU-E, CFU-GM) were performed on 2×10^5 cell aliquots of transduced cells before cryopreservation to assess their viability as well as to determine the pretransplant MDR transduction efficiency.

Surveillance for the Presence of Replication-Competent Retrovirus

One percent of the transduced cells and 5% of the suspension media was tested by the National Gene Vector Laboratory (NGVL) in Indianapolis for the presence of replication-competent retrovirus (RCR). This material, as well as samples of blood and marrow harvested from the patients at 2 weeks, 4 weeks, and 3, 6, and 9 months post transplant (when available), was assayed by coculture with *Mus dunni* cells for 4 weeks, followed by PG4 S + L-assay [23]. A portion of the cell media was processed for Gram staining and bacterial and fungal culture to monitor for contamination during the 72- to 96-h incubation period.

Results

Patient Characteristics

Seven patients were entered in this study. Prior therapy and posttransplant outcome of these patients is detailed in Table 1. The single patient with a brain tumor had been treated with surgery and whole-brain radiation therapy after

Table 1. Patient characteristics and outcome

Patient	Age/sex	Diagnosis	Date of diagnosis	Prior chemotherapy	Date of BMT	Outcome
1	47/M	Glioblastoma	10/94	BCNU ×3	08/02/95	Died of disease 10/96
2	44/F	Breast	4/90	CAF ×6, paclitaxel/cisplatin ×6	10/03/95	Progressed after registration; removed from study; died of disease 06/96
3	31/F	Breast	4/94	CAF ×4, paclitaxel ×2, paclitaxel/cisplatin ×6, 5FU/leucovorin ×4	11/13/95	Progressed after registration; removed from study; died of BMT complications 12/95
4	53/F	Breast	4/91	CAF ×6, paclitaxel ×6, Mit-C/velban ×4, 5FU/leucovorin ×6	12/28/95	NED
5	46/F	Breast	1/95	CAF ×6	02/02/96	NED
6	44/F	Ovary	11/90	Cyclophosphamide/cisplatin ×6, carboplatin ×6, paclitaxel ×6, daunorubicin/etoposide ×6	02/22/96	Died of disease 09/96
7	47/F	Ovary	06/95	Paclitaxel/cisplatin ×7	7/24/96	Died of complications 08/96

NED, no evaluable disease; CAF, cyclophosphamide/Adriamycin/5-FU; BMT, blood or marrow transplant.

diagnosis and received BCNU (standard chemotherapy for patients with brain tumors) while awaiting insurance approval to proceed with high-dose chemotherapy. Of the four patients with breast cancer, patient 2 had pleural-based metastatic disease diagnosed 4 years after initial adjuvant CAF chemotherapy, while patient 3 presented with inflammatory breast cancer that spread locally while she received CAF chemotherapy, and progressed to involve the lungs on paclitaxel-based chemotherapy. She ultimately had a partial response to 5-fluorouracil/leucovorin. Patient 4, 3 years after initial adjuvant CAF, developed liver metastases that responded partially to weekly 5-fluorouracil/leucovorin. Patient 5 presented with lung metastases that almost completely resolved with standard CAF chemotherapy. Both patients 6 and 7 with ovarian cancer had stage IV disease. Patient 6 had received multiple regimens of chemotherapy since the presentation of her disease with only a short disease-free interval after her initial response to chemotherapy in 1991. Patient 7, who had extensive abdominal and pulmonary disease, had a complete response to induction therapy.

Patients 2 and 3 progressed after registration and marrow harvesting and were removed from the gene therapy study, patient 2 because of the presence of metastatic disease noted in the marrow harvest and patient 3 for bone metastases. These two patients underwent high-dose therapy but received only untransduced cells for hematopoietic support. All other patients underwent high-dose therapy with reinfusion of the transduced stem cells. Toxicities were those expected for dose-intensive therapy requiring stem cell support (pancytopenia, fever, diarrhea, mucositis, multiorgan failure). No toxicities specifically attributable to the use of transduced stem cells were observed. Length of hospital stay, time to engraftment, and blood product requirements post transplantation are presented in Table 2. Time to engraftment for patient 1 was prolonged as he had received carmustine (BCNU), a stem cell poison, for his brain tumor, before marrow harvesting. It is notable that patient 3, who was also heavily pretreated with drugs including cisplatin before undergoing her course of high-dose chemotherapy, had a protracted period of pancytopenia although she did not receive any transduced cells.

Evaluation for Exogenous MDR Gene Transfer

Marrow was used for transduction for the first six patients entered on this study, although both marrow- and blood-derived stem cells were reinfused to ensure prompt engraftment. After the protocol was amended, only blood-derived stem cells were harvested from patient 7, a portion of these being transduced and the remainder cryopreserved unmanipulated. PCR of marrow of patients post transduction and before transplantation showed that, even

TABLE 2. Engraftment and transfusion requirements

Patient	Regimen	Days in hospital	Days to ANC >500/µl	Days to platelet count >20 000/µl	Units transfused Blood	Units transfused Platelets
1	CAMP 4	75	30	70	24	171
2[a]	CAMP 1	21	8	12	5	18
3[a]	CAMP 1	49	25	Not achieved	39	214
4	CAMP 1	28	9	28	5	60
5	CAMP 7	18	9	11	2	24
6	CAMP 3	20	8	30	4	72
7	CAMP 3	22	9	15	20	90

ANC, absolute neutrophil count; CAMP 1, cyclophosphamide $6 g/m^2$, thiotepa $500 mg/m^2$, carboplatin $800 mg/m^2$; CAMP 3, ifosfamide $1 500 mg/m^2$, thiotepa $300 mg/m^2$, carboplatin $1 500 mg/m^2$; CAMP 4, BCNU $450 mg/m^2$, thiotepa $500 mg/m^2$, etoposide $1 200 mg/m^2$; CAMP 7, tandem 3-cycle high-dose regimen of paclitaxel $500–825 mg/m^2$ followed by melphalan $180 mg/m^2$, followed by the CAMP 1 regimen.
[a] Patients who did not receive transduced cells.

with the scale-up of retroviral supernatant and CD34+ cells used, 20%–70% of BFU-E and CFU-GM derived from transduced CD34+ cells was positive by MDR PCR (Table 3). PCR analysis of patient marrow samples post transplantation for the transferred exogenous MDR cDNA were positive in one sample each from patients 1 and 7 (Table 3). Semiquantitation of the MDR PCR signal with control samples of A12M1 diluted to varying extents revealed that fewer than 1 in 1 000 cells in the sample were transduced. PCR positivity was observed in the two patients reinfused with the highest percentage of MDR-transduced cells, 25% and 26%, respectively as compared to lesser percents in the other patients (Table 3). FACS analysis of patient samples for increased p-glycoprotein expression and methylcellulose analysis of individual BFU-E and CFU-GM by MDR PCR have all been negative. In addition, the exposure of MDR-transduced cells from patient 1 at 10 weeks post transplant to $1 \mu g/ml$ paclitaxel in vitro for 6 and 24 h demonstrated no evidence of p-glycoprotein expression. Patients 4 and 5 have not exhibited any disease progression, making further chemotherapy unnecessary at present, and are continuing to be monitored for MDR gene transfer and expression.

Surveillance for the Presence of Replication-Competent Retrovirus

As indicated earlier, 1% of the transduced cells and 5% of the suspension media was collected and evaluated by the National Gene Vector Laboratory (NGVL) in Indianapolis for the presence of replication-competent retrovirus (RCR). This material, as well as samples of blood and marrow harvested from the patients at 2 weeks, 4 weeks, and 3, 6, and 9 months post transplant (when

TABLE 3. Transduction efficiency and posttransplant MDR-PCR results

Patient	Transduction method	Transduced	Positive colonies (%)		Transduced cells/cells infused	Post BMT PCR
			BFU-E	CFU-GM		
1	AIS plates	Marrow	30	50	25%	+10 weeks
2	CellPro Column	Marrow	NA	NA	Progressed after registration; removed from study	
3	CellPro Column	Marrow	33	25	Progressed after registration; removed from study	
4	CellPro Column	Marrow	40	20	10%	Up to 8 months
5	CellPro Column	Marrow	56	40	14%	Up to 6 months
6	CellPro Column	Marrow	70	33	18%	Up to 6 months
7	CellPro Column	Blood	NA	NA	26%	+3 weeks

available), was assayed by coculture with *Mus dunni* cells for 4 weeks, followed by PG4 S + L-assay, and all samples have thus far been negative for RCR in this sensitive assay.

Discussion

MDR expression of *p*-glycoprotein, the transmembrane protein product of MDR in committed progenitor cells, is insufficient to prevent marrow toxicity in human marrow exposed to MDR-sensitive drugs. Hematopoietic stem cells cells have some measurable MDR gene expression and thus may be relatively protected from the cytotoxic effects of natural products. Despite the fact that some of the MDR-sensitive drugs have dose-limiting nonhematological toxicities, myelosuppression by these drugs often results in significant morbidity and mortality, and limits the amount of drug that can be used. Repetitive doses of combinations of MDR drugs given at higher than conventional doses may be feasible in patients with MDR-transduced hematopoietic progenitors, protected by increased *p*-glycoprotein expression. These intensive drug doses may kill additional tumor cells, resulting in a higher percentage of complete responses or more durable complete responses.

We and others have shown that transduction and expression in HPC of an exogenous MDR gene using a strong promoter results in the selection of resistant marrow stem cells on exposure to MDR-affected drugs in mice [7,8,10]. This finding provides a rational approach to protecting human bone marrow cells from the toxicity of these drugs. In culture, we have achieved high-level MDR transduction and *p*-glycoprotein expression of both bone marrow- and peripheral blood-derived human HPC using fibronectin-coated plates with a protocol in which CD34+ cells are subjected to a 48-h preincubation with growth factors IL-3, IL-6, and stem cell factor (SCF), and then to a 24-h incubation with MDR-cDNA containing retroviral supernatants [9,11].

The primary aim of this study was to transduce CD34+ hematopoietic cells of patients with advanced breast, ovarian, or brain tumors with a safe and effective retrovirus containing and expressing the MDR gene, and to use these MDR-transduced cells to provide these patients with a population of cells potentially resistant to the myeloablative effects of anticancer chemotherapy. As this was a phase I study of gene therapy, a number of safeguards were incorporated. Eligibility criteria included only patients with advanced disease for whom no curative therapy existed. Predictably, therefore, two patients progressed after study entry despite continued induction treatment and were no longer eligible. To decrease the risk of gene transduction and infusion of contaminating cancer cells in this study, bilateral marrow samples were required to be negative for tumor before marrow harvest, patients with bone metastases (at higher risk for marrow involvement) were ineligible, and the protocol specifically included patients with ovarian and brain primaries in which marrow involvement is uncommon. Finally, stem cells were selected for CD34 positivity before the gene transduction process to improve the ratio of the number of retroviral particles compared to the number of CD34+ cells to nearly 1.

The protocol was designed to transduce only one-third of the reinfused stem cells to avoid the risk of failure of marrow recovery and to ensure reconstitution with adequate numbers of unmanipulated stem and progenitor cells. Marrow was used initially rather than blood-derived stem cells because clinical experience with blood-derived stem cells at the time of the initiation of the study had not yet documented long-term engraftment of these cells, and laboratory studies had not defined techniques for efficient transduction of blood-derived stem cells. Laboratory and clinical data now suggest that blood-derived stem cells are more easily and readily available than marrow cells, are at least as efficiently transduced, and are, thus, the cells of choice for human HPC retroviral transduction [11,24].

The specific potential risk of the retroviral gene transduction procedure is the creation of a productive retroviral infection by a recombination of the packaging cell line genome with the inserted gene. The transfer of such a recombinant retrovirus to the bone marrow cells could affect the normal engraftment potential of the marrow. To avoid this problem, we used a retroviral packaging line, GPAm12, which contains multiple safeguards against the generation of replication-competent retrovirus (RCR), as compared to other available packaging lines [13]. To monitor for the possible generation of RCR, we assayed for the presence of intact retrovirus in the marrow- or blood-derived stem cells of the patients at the time of reinfusion, as well as at specific time points after successful engraftment. No evidence of retroviral activity was detected in these assays.

A further aspect of this protocol was to determine any other toxicities potentially attributable to gene transduction as well as any delay in engraftment from manipulation of one-third of the reinfused stem cells. We anticipated and observed complications unavoidable in patients undergoing high-dose chemotherapy. Patients were carefully evaluated for any unexpected adverse events. No toxicities attributable to stem cell transduction were observed. Two patients (one who received transduced cells and one who did not) died during the transplant process. Only patient 1 had slow engraftment, attributable to his prior treatment with the stem cell toxin, carmustine, rather than the transduction procedure. Patient 3, who did not receive transduced cells, did not achieve platelet engraftment before her death. The median duration of hospitalization as well as the blood and platelet transfusion requirements were similar for this group of patients with advanced, heavily pretreated cancer whether they received gene-transduced cells or not.

Finally, we were interested in the success of gene transfer as assessed by evaluating the marrow post transplantation by MDR PCR analysis. We achieved high levels of MDR gene transduction in PCR analysis of samples in the scaled-up CD34+ cells eventually given to patients before their reinfusion: 30%–70% of BFU-E and 20%–50% of CFU-GM colonies were transduced (see Table 3). These levels were almost as high as in small-scale experiments in which we showed 50%–60% transduction of BFU-E and CFU-GM derived from CD34+ cells using the same transduction protocol as in this trial [11]. In that study, we also showed that approximately 25% of the transduced clones were resistant to a dose of taxol (10^{-8} M) that killed more than 95% of untransduced controls. The results of this clinical trial indicate that the use of large volumes of MDR A12M1 supernatant and CD34+ cells in the clinical setting retains the ability to achieve high-level MDR transduction of human HPC. These levels of MDR transduction are much higher than in a recent report using other procedures [25]. However, despite the high level of pretransplant MDR transduction, only two posttransplant marrow samples, from patients 1 and 7, were positive for an exogenous MDR signal by PCR. In addition, quantitation of the samples compared to dilutions of the A12M1 clone showed that fewer than 1 in 1 000 cells were transduced. Further treatment with MDR-sensitive drugs might be expected to select for preferential survival of MDR-transduced cells and result in PCR positivity in the remaining patients receiving transduced cells. Two patients receiving transduced cells remain without disease progression at present, and, although no treatment is currently indicated, they may receive taxol or other therapy if they relapse and thus test this selection hypothesis in vivo.

We have previously shown that preincubation of CD34+ cells with IL3, IL6, and SCF and exposure to these growth factors during retroviral transduction is essential for optimizing transduction as it allows the cells to divide, a requirement for integration of the transduced gene [11]. It is unlikely that treatment with the IL-3, IL-6, and SCF cytokine mix prevents stem cell or progenitor transduction and MDR expression as we have previously shown long-term high-level MDR transfer and expression in marrow-ablated mice [7,22]. In addition, we have shown significant numbers of transduced cells derived from CD34+ cells in culture using both methylcellulose assay of progenitors and the long-term culture-intiating cell (LTC-IC) assay, believed to be the best surrogate assay for human stem cells [9,11]. However, there are no competing unmanipulated and untransduced cells in these models. We believe that the major reason for our poor results of MDR transduction of reinfused transduced cells post transplantation in this clinical trial is the unfavorable competition of the cytokine-stimulated transduced cells with unmanipulated untransduced cells that were given simultaneously to ensure marrow reconstitution in these patients.

Competition from the untransduced cells may be enhanced by the cytokines used in the transduction protocol, which allow HPC to be driven to either differentiate or otherwise change their biology so that they no longer compete favorably with untransduced cells. This relative engraftment defect of IL-3-, IL-6-, and SCF-stimulated cells has been demonstrated in competition studies in mice [26]. IL-3 stimulation, in particular, appears to reduce the long-term repopulating ability of stem cells [27,28]. By contrast, other newer combinations of cytokines containing IL-1, IL-6, SCF, IL-11, Flt3, and thrombopoietin appear not to alter the repopulating ability of stimulated HPC [27–32]. The use of these newer growth factor combinations may allow transduced cells to compete more favorably or even equally with unmanipulated cells for marrow reconstitution, and are now being tested in preclinical settings for potential use in future protocols. Because both the patients who successfully engrafted with detectable transduced cells received a higher percentage of transduced as compared to untransduced progenitor cells (25% and 26%) than those who were PCR negative, it is possible that the negative findings may also result from the reinfusion of inadequate numbers of transduced HPC. Certainly this latter hypothesis has been strengthened by recent reports from our laboratory, which showed that by increasing the percentage of male cells infused together with female cells (the total being sufficient to allow for engraftment) into lethally irradiated female mice, the degree of "maleness" of the female recipients can be increased by increasing the percentage of male cells as compared with female cells infused at the same time point. In addition, the infusion of the male cells 48h before the infusion of the female cells also provides an improved engraftment potential to the male cells [33].

In summary, in this phase 1 trial, we have demonstrated the safety of the transduction process itself with respect to normal marrow engraftment, and have shown that we can attain high-level MDR transduction of colonies derived from CD34+ cells after large-scale, clinically applicable, protocol preparation. Further studies in patients will be continued using a new cytokine "cocktail" comprising SCF, MGDF, and G-CSF. Furthermore, a minimum number of transduced cells (30%) will be infused together with, or 48h before, the untransduced cells to confirm that competition was a major reason for the lack of demonstration of MDR-transduced cells in the patients receiving proportionately low numbers of transduced cells.

Acknowledgments. This investigation was supported in part by the following grant numbers awarded by the National Cancer Institute, DHHS: NCI UOI CA65838 MDR Gene Therapy for Drug Resistance in Breast Cancer, NCI R21

156   C. Hesdorffer et al.

CA66244-01; Columbia Cancer Center Breast Cancer Research Programs; P30-
CA13696-21 to support the Herbert Irving Comprehensive Cancer Center.
A.B. has an equity interest in Genetix Pharmaceuticals Inc.

References

1. Frei E III, Antman K, Teicher B, et al (1989) Bone marrow autotransplantation for solid tumors-prospects. J Clin Oncol 7:515–526
2. Hryniak W, Bush H (1984) The importance of dose intensity in chemotherapy of metastatic breast cancer. J Clin Oncol 2:1281–1288
3. DeVita VT, Hubbard SM, Young RC, et al (1988) The role of chemotherapy in diffuse aggressive lymphomas. Semin Hematol 25:2–10
4. Bezwoda WR, Seymour L, Dansey RD (1995) High-dose chemotherapy with hematopoietic rescue as primary treatment for metastatic breast cancer: a randomized trial (see comments). J Clin Oncol 13:2483–2489
5. Attal M, Harousseau JL, Stoppa AM, et al (1996) A prospective, randomized trial of autologous bone marrow transplantation and chemotherapy in multiple myeloma. Intergroupe Francais du Myelome. N Engl J Med 335:91–97
6. Martelli M, Vignetti M, Zinzani PL, et al (1996) High-dose chemotherapy followed by autologous bone marrow transplantation versus dexamethasone, cisplatin, and cytarabine in aggressive non-Hodgkin's lymphoma with partial response to front-line chemotherapy: a prospective randomized Italian multicenter study. J Clin Oncol 14:534–542
7. Podda S, Ward M, Himelstein A, et al (1992) Transfer and expression of the human multiple drug resistance gene into live mice. Proc Natl Acad Sci USA 89:9676–9680
8. Sorrentino BP, Brandt SJ, Bodino G, et al (1992) Selection of drug-resistant bone marrow cells in vivo after retroviral transfer of human MDR1. Science 257:99–103
9. Ward M, Richardson C, Pioli P, et al (1994) Transfer and expression of the human multiple drug resistance gene in human CD34+ cells. Blood 84:1408–1414
10. Hanania E, Deisseroth AB (1994) Serial transplantation shows that early hematopoietic precursor cells are transduced by MDR-1 retroviral vector in mouse gene therapy model. Cancer Gene Ther 1:21–25
11. Ward M, Richardson C, Pioli P, et al (1996) Retroviral transfer and expression of the human MDR gene in peripheral blood progenitor cells. Clin Cancer Res 2:873–876
12. Hesdorffer C, Antman K, Bank A, et al (1994) Clinical protocol: human MDR gene transfer in patients with advanced cancer. Hum Gene Ther 5:1151–1160
13. Markowitz D, Goff S, Bank A (1988) Construction and use of a safe and efficient amphotropic packaging cell line. Virology 167:400–405
14. Pastan I, Gottesman MM, Ueda K, et al (1988) A retrovirus carrying an MDR cDNA confers multidrug resistance and polarized expression of P-glycoprotein. Proc Natl Acad Sci USA 85:4486–4490
15. Lebkowski JS, Schain LR, Okrongly D, et al (1992) Rapid isolation of human CD34 hematopoietic stem cells—purging of human tumor cells. Transplantation 53:1011–1019
16. Okarma T, Lebkowski J, Schain L, et al (1992) The AIS selector: a new technology for stem cell purification. In: Advances in bone marrow purging and processing. Wiley-Liss, New York, pp 487–504

17. Berenson RJ, Bensinger WI, Kalamasz DF, et al (1992) Transplantation of stem cells enriched by immunoadsorption. Prog Clin Biol Res 377:449–457

18. Berenson RJ (1992) Transplantation of CD34+ hematopoietic precursors: clinical rationale. Transplant Proc 24:3032–3034

19. Heimfeld S, Fogarty B, McGuire K, et al (1992) Peripheral blood stem cell mobilization after stem cell factor or G-CSF treatment: rapid enrichment for stem and progenitor cells using the CEPRATE immunoaffinity separation system. Transplant Proc 24:2818

20. Noonan KE, Beck C, Holzmayer TA, et al (1990) Quantitative analysis of MDR1 (multidrug resistance) gene expression in human tumors by polymerase chain reaction. Proc Natl Acad Sci USA 87:7160–7164

21. Aihara M, Aihara Y, Schmidt-Wolf G, et al (1991) A combined approach for purging multidrug-resistant leukemic cell lines in bone marrow using a monoclonal antibody and chemotherapy. Blood 77:2079–2084

22. Richardson C, Bank A (1995) Preselection of transduced murine hematopoietic stem cell populations leads to increased long-term stability and expression of the human multiple drug resistance gene. Blood 86:2579–2589

23. Lander MR, Chattopadhyay SK (1984) A *Mus dunni* cell line that lacks sequences closely related to endogenous murine leukemia viruses and can be infected by ectropic, amphotropic, xenotropic, and mink cell focus-forming viruses. J Virol 52:695–698

24. Bregni M, Magni M, Siena S, et al (1992) Human peripheral blood hematopoietic progenitors are optimal targets of retroviral-mediated gene transfer. Blood 80:1418–1422

25. Hanania EG, Giles RE, Kavanagh J, et al (1996) Results of MDR-1 vector modification trial indicate that granulocyte-macrophage colony-forming unit cells do not contribute to posttransplant hematopoietic recovery following intensive chemotherapy. Proc Natl Acad Sci USA 93:15346–15351

26. Peters SO, Kittler EL, Ramshaw HS, et al (1996) Ex vivo expansion of murine marrow cells with interleukin-3 (IL-3), IL-6, IL-11, and stem cell factor leads to impaired engraftment in irradiated hosts. Blood 87:30–37

27. Zandstra PW, Conneally E, Piret JM, et al (1996) Normal bone marrow cells capable of generating expanded populations of LTC-IC are CD34+CD38–, require particularly high concentrations of flt3-ligand (FL) and are inhibited by excess IL-3. Blood 88:445a

28. Yonemura Y, Ku H, Lyman SD, et al (1997) In vitro expansion of hematopoietic progenitors and maintenance of stem cells: comparison between flt3/flk2 ligand and kit ligand. Blood 89:1915–1921

29. Trevisan M, Yan XQ, Iscove NN (1996) Cycle intiation and colony formation in culture by murine marrow cells with long-term reconstituting potential in vivo. Blood 88:4149–4158

30. Borge OJ, Ramsfjell V, Veiby OP, et al (1996) Thromobpoietin, but not erythropoietin promotes viability and inhibits apoptosis of multipotent murine hematopoietic progenitor cells in vitro. Blood 88:2859–2870

31. Conneally E, Cashman J, Petzer AL, et al (1996) In vitro expansion of human lymphomycloid stem cells from cord blood demonstrated using a quantitative in vivo repopulating assay. Blood 88:628a

32. Conneally E, Eaves CJ, Humphries RK (1996) High efficiency gene transfer to primitive human hematopoietic stem cells capable of multilineage repopulation of immunodeficient NOD/SCID mice. Blood 88:646a

33. Qin S, Ward M, Raftopoulos H, Hesdorffer C, Bank A (1998) Delayed infusion of normal hematopoietic stem cells increases marrow reconstitution of retrovirally-transduced cells. Blood 90:118a

Basic Studies Toward Hematopoietic Stem Cell Gene Therapy

Yutaka Hanazono[1], Cynthia E. Dunbar[2], Robert E. Donahue[2],
Ikunoshin Kato[3], Yasuji Ueda[4], Mamoru Hasegawa[4], Masashi Urabe[1],
Akihiro Kume[1], Keiji Terao[5], and Keiya Ozawa[1]

Summary. Hematopoietic stem cells (HSCs), because they have a self-renewal ability and can generate progeny of all kinds of blood cells throughout one's life, are an ideal target for gene therapy. Retroviral vectors are predominantly used for transduction of HSCs, but the gene transfer efficiency is extremely low. Several efforts have been made at achieving clinically relevant gene transfer efficiencies. First, new cytokines such as Flt-3 ligand and thrombopoietin, and coculture with stromal elements such as fibronectin fragments, have been successfully tried during ex vivo culture of HSCs with retroviral vectors. Second, new vectors that meet the host requirements have been developed: pseudotyped retroviral vectors and lentiviral vectors. Finally, positive selection of transduced cells has been designed in vitro before reinfusion or in vivo after engraftment to compensate for the low transduction efficiency of HSCs. A novel method of in vivo expansion of transduced hematopoietic cells using the selective amplifier gene may also help overcome the low transduction efficiency of HSCs. It has recently been reported that immunological tolerance against xenogeneic gene products can be induced by introduction of their genes into HSCs. This distinctive feature further enhances the value of HSCs as a target of gene therapy.

Key words. Hematopoietic stem cell, Gene therapy, Retroviral vector, Selective amplifier gene, Tolerance

[1] Division of Genetic Therapeutics, Center for Molecular Medicine, Jichi Medical School, 3311-1 Yakushiji, Minamikawachi-machi, Kawachi-gun, Tochigi 329-0498, Japan
[2] Hematology Branch, National Heart, Lung and Blood Institute, National Institutes of Health, 9000 Rockville Pike, Bethesda, MD 20892, USA
[3] Takara Shuzo Co., Ltd., 3-4-1 Seta, Otsu, Shiga 520-2193, Japan
[4] DNAVEC Research Inc., 1-25-11 Kannondai, Tsukuba, Ibaraki 305-0856, Japan
[5] Tsukuba Primate Center for Medical Science, National Institute of Infectious Diseases, 1 Yahatadai, Tsukuba, Ibaraki 305-0843, Japan

Introduction

Hematopoietic stem cells (HSCs) have an ability of self-renewal and can generate progeny of all kinds of blood cells throughout one's life. Therefore, HSCs have been considered to be an ideal target for gene therapy ever since technology was developed allowing gene transfer into eukaryotic cells [1]. The clinical applications of stem cell gene therapy include three major categories at present: first, replacement of a missing or damaged gene product; second, chemoprotection during anticancer therapy by transfer of drug resistance genes to normal HSCs; and third, intracellular vaccination against HIV infection and cancer.

Although several kinds of vectors have been studied for gene therapy applications, retroviral vectors are predominantly used for gene transfer to HSCs (Table 1). Vectors that can integrate into a host genome are required for transduction of HSCs. Retroviral vectors are the only vector that can integrate efficiently into host genome to date, although another integrating vector, the lentiviral vector, is being developed [2].

Gene Transfer into Hematopoietic Stem Cells

Animal models are essential for evaluating gene transfer into HSCs, because at present there are no reliable in vitro assays for cells capable of constituting long-term hematopoiesis in vivo. Successful transduction of murine HSCs

TABLE 1. Vectors used for gene transfer

Vector	Retrovirus	Adeno-associated virus	Adenovirus	Plasmid DNA	HIV
Stem cell transduction efficiency	Low—moderate (cf. Table 2)	Low	Low	Extremely low	Low—moderate? (pseudotype required)
Integration into host genome	Yes	Inefficient	No	Very inefficient	Yes
Replication-incompetent vectors	Yes	Yes?	Yes?	Yes	Yes?
Dependency on cell cycle	Yes	No	No	No	No
Clinical applications	Yes	Yes	Yes	Yes	No

TABLE 2. Efficiencies of gene transfer into hematopoietic cells with retroviral vectors

Cells	Mice	Monkeys	Humans
Progenitor cells (in vitro transduction efficiency of colony-forming cells)[a]	~50%	~50%	~50%
Stem cells (in vivo transduction efficiency after transplantation of cells)[b]	~50%	<1%	<1%

[a] Cells after transduction were plated in methylcellulose to form colonies, and DNA from each colony was assessed for a vector sequence.
[b] After myeloablation and transplantation of transduced cells, DNA from peripheral blood cells was assessed for a vector sequence.

with retroviral vectors derived from mouse retroviruses has now been carried out almost routinely by many laboratories, since it was first reported in 1984 [3]. However, stem cell-targeted gene transfer efficiency in humans and monkeys was found to be extremely low (usually <1%) in contrast to the results obtained in murine studies (about 50%), although hematopoietic progenitor-targeted gene transfer efficiencies are similar among these species (Table 2) [4].

There are two possible explanations for this discrepancy. First, cell cycle kinetics are different between primate and murine HSCs. The self-renewal frequency of murine HSCs is presumably high, so that cells are expected to divide at least once during a few days of a standard ex vivo transduction period. However, in larger animals, the frequency of self-renewal of HSCs may be much lower, even in the presence of hematopoietic cytokines. It is reported that feline HSCs divide at most once every 3 weeks [5]. It is therefore unlikely that these cells cycle during a transduction period.

Second, most murine experiments have used retroviral vectors packaged in an ecotropic envelope protein that binds to a receptor not found on primate cells. Instead, transduction of primate cells must utilize vector particles incorporating an amphotropic envelope protein [6]. The most primitive rhesus monkey and human hematopoietic cells (CD34+/CD38−) appear to have extremely few amphotropic receptors, whereas primitive murine stem cells have more ecotropic receptors [7]. The improvement of transduction efficiency by cytokines may be caused, at least partially, by upregulation of retroviral receptors on the surface of target cells [8].

Optimizing Transduction Methods

Future promising approaches for improvement of HSC transduction efficiencies are summarized in Table 3. Cytokines such as Flt-3 ligand and thrombopoietin that are shown to have activity on very primitive hematopoietic

TABLE 3. Optimization for gene transfer to human hematopoietic stem cells

Modification of ex vivo culture	Introduction of new cytokines: Flt-3 ligand Thrombopoietin Coculture with stromal elements: Fibronectin fragment (CH-296) Abrogation of negative regulators: Anti-TGF-β antibody Anti-p27^{kip-1} oligonucleotide
Development of vectors	Pseudotype: Vesicular stomatitis virus (VSV)-G envelope Gibbon ape leukemia virus (GALV) envelope Lentivirus: HIV vectors
Selection of transduced HSCs	Ex vivo: CD24 Green fluorescent protein (GFP) Truncated low-affinity nerve growth factor receptor (tLNGFR) In vivo: Multidrug resistance-1 (MDR-1) Dihydrofolate reductase (DHFR) Selective amplifier gene

cells have been introduced into ex vivo culture with retroviral vectors [9–13]. Coculture with specific stromal elements such as fibronectin fragments has been successfully tried [14]. Abrogation of negative regulators of hematopoiesis such as transforming growth factor-β (TGF-β) and p27^{KIP-1} can be considered [15,16]. A few recent papers reported that, with the addition of Flt-3 ligand and fibronectin fragments, clinically relevant levels of greater than 10% in vivo gene transfer were obtained in rhesus and baboon models [17,18].

Alternative vector systems are also being developed. Pseudotyping, or replacement of the amphotropic envelope protein with an alternative protein, has been developed to change target cell specificity or the physical properties of the vector particle. The presence of the vesicular stomatitis virus (VSV)-G envelope protein results in increased viral stability and allows concentration of vector preparations [19]. Pseudotyping with gibbon ape leukemia virus (GALV) envelope protein improves the transduction efficiency of primate lymphocytes [20], and possibly HSCs [21], when compared to standard amphotropic vectors.

Lentiviruses such as HIV do not require mitosis for transduction and integration because they have nuclear translocation signals. HIV-based vectors are reported to be able to transduce nondividing cell populations [2]. Pseudotyping of HIV-based vectors with VSV-G or amphotropic envelope protein is needed for transduction of HSCs because CD4, the receptor for HIV, is not present on repopulating stem cells. A preliminary study of a pseudotyped HIV vector demonstrated significant improvement of gene transfer into a quiescent population of primitive myeloid progenitors, implying efficient human HSC transduction [22]. For clinical applications, however, establishment of an efficient stable vector-producing system is required. Another more serious limitation of HIV vectors is the remote possibility that they could recombine into virulent forms and cause AIDS, although it may be reduced to a negligible risk.

Ex Vivo Selection of Transduced Stem Cells

Many recent papers describe so-called ex vivo expansion of human progenitor and stem cells, either in suspension culture with cytokines or in bioreactors or other devices containing a stromal support layer [23]. Most studies report increases in total cell numbers, colony-forming units, or total CD34+ cells, probably inadequate surrogates for true stem cells, just as transduction of these cell types does not predict transduction of true stem cells. Gene marking studies will help clarify the kinetics of hematopoiesis originating from these ex vivo expanded populations. Gene marking studies in rhesus monkeys indicate that ex vivo expansion of mobilized peripheral blood cells for 10–14 days in the presence of interleukin (IL) -3, IL-6, stem cell factor, Flt-3 ligand, and the fibronectin fragment CH-296 results in no increase in initial engraftment and diminished long-term engraftment [17]. A recent report of graft failure after transplantation of ex vivo expanded cells supports the conclusion that the present "expansion" conditions may damage engrafting cells and that committed progenitors do not contribute to even short-term engraftment [24].

Low efficiency of gene transfer into HSCs may be compensated for by positive selection of transduced cells in vitro before reinfusion or in vivo after engraftment. Drug selection of primary cells with neomycin requires ex vivo culture for days to weeks, likely changing the characteristics of target cells through terminal differentiation or other processes. More rapid selection of transduced cells can be performed using marker genes encoding proteins detectable by fluorescence-activated cell sorting (FACS).

The human cell-surface protein CD24 or its murine analog, heat-stable antigen, has been tested as a selectable marker. Successful separation of transduced cells from untransduced cells has been reported by FACS based on cell-surface expression of CD24 [25]. However, of practical concern, CD24 and HSA are glycosylphophatidylinositol-linked cell-surface glycoproteins, and cell–cell transfer of this class of proteins can occur, abrogating specificity and potentially confusing marking interpretation [26].

Jellyfish green fluorescent protein (GFP), a naturally fluorescent cytoplasmic protein, has been developed as a marker gene. Specific wavelength-induced fluorescence of GFP does not require cofactors, substrates, or additional gene products. The signal is detectable by fluorescent microscopy or FACS in live cells [27]. Derivation of stable retroviral producer cell lines expressing GFP has been difficult [28]. Recently, GFP vectors have been successfully used to transduce murine and rhesus monkey hematopoietic stem cells and to track gene-modified cells in vivo [29–31]. However, there is a concern that GFP may elicit a host immune response, resulting in elimination of transduced cells.

A truncated form of the human low-affinity nerve growth factor receptor (LNGFR) is also in development as a selectable marker for hematopoietic cells [32]. LNGFR is normally not present on the surface of hematopoietic cells from any lineage. The human LNGFR gene sequences can be used as a unique marker tag in these lineages, eliminating complications arising from immunogenicity of the transgene. To prevent signal transduction through this receptor in hematopoietic cells, the coding sequences for the cytoplasmic transducing domains of the LNGFR gene were deleted. Vectors containing this transgene can transduce lymphocytes efficiently, and selection of transduced cells using FACS for the truncated LNGFR is very sensitive and specific [33]. However, there is a concern regarding the ectopic expression of a receptor protein capable of ligand binding.

In Vivo Selection of Transduced Stem Cells

In vivo selectable drug resistance genes have also been incorporated into retroviral vectors. In mouse models, the multidrug resistance (MDR-1) gene allows for in vivo enhancement of the proportion of hematopoietic cells containing a vector, with implications for correction of genetic disorders requiring high percentages of corrected cells [34]. Engraftment with hematopoietic stem cells carrying drug resistance genes may confer chemoprotection and thus permit dose intensification in patients being treated for nonhematopoietic tumors. Several trials completed or in progress have transplanted HSCs transduced with vectors containing the MDR-1 gene, followed by chemotherapy with an MDR-pumped drug such as taxol. However, it has recently been reported that ex vivo expanded murine bone marrow cells transduced with an MDR-1 vector cause a myeloproliferative syndrome in transplanted mice [35]. A number of alternative drug resistance genes have also been studied in vitro and in murine models including the antifolate-resistant dihydrofolate reductase [36].

In vivo selection of transduced cells using drug resistance genes requires administration of anticancer drugs, which is a drawback to treatment of

patients with nonmalignant diseases. To overcome this disadvantage, we constructed a retroviral vector that confers in vivo selective expansion of transduced cells. The vector expresses a chimeric cDNA as a selective amplifier
gene that encodes the fusion protein consisting of the signal-transducing
domain of the granulocyte colony-stimulating factor (G-CSF) receptor and
the hormone-binding domain of the estrogen receptor, allowing to activate
the exogenous G-CSF receptor by treatment with estrogen (Fig. 1) [37]. It is
worthy of note that erythroid colonies were formed from the bone marrow
cells transduced with this chimeric gene in the presence of estrogen without
the addition of erythropoietin, suggesting that the signal from the G-CSF
receptor portion of the chimeric molecule does not preferentially induce neu-
trophilic differentiation, but promotes the differentiation depending on the
nature of the target cells.

Furthermore, the chimeric protein had Tyr703 of the G-CSF receptor
mutated to Phe, because this residue plays a pivotal role in transmitting the
differentiation signal. When the mutated chimeric gene was introduced into
IL-3-dependent cell line 32D cells that undergo granulocytic differentiation

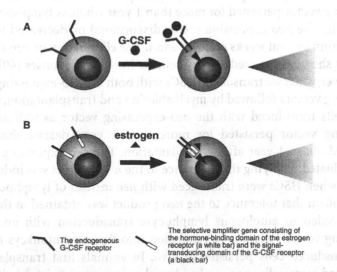

FIG. 1A,B. Expansion of hematopoietic stem/progenitor cells transduced with the selective amplifier gene. A When cells that have the intrinsic granulocyte colony-stimulating
factor (G-CSF) receptor are treated with G-CSF, the G-CSF receptors are dimerized, allowing to transmit the signal and the cells proliferate or differentiate. B When cells are transduced with the selective amplifier gene and are treated with estrogen, the chimeric
receptors are dimerized, allowing the signal to transmit, and the transduced cells eventually proliferate. To avoid transmitting the differentiation signal, Tyr703 of the G-CSF receptor was mutated to Phe

upon G-CSF treatment, the transduced 32D cells continued to proliferate in response to estrogen without granulocytic differentiation. This result suggests that the mutated chimeric gene predominantly transmits growth signals without inducing obvious differentiation, and thus the modified gene would function as a selective amplifier gene for hematopoietic stem cell gene therapy. We are planning to examine in vivo effects of the gene in a cynomolgus monkey HSC transplantation model.

Immunological Characteristics of Stem Cell Gene Transfer

Introduction of xenogeneic genes has been found to elicit strong immune responses that eliminate transduced cells such as lymphocytes [38]. To study immune responses to transduced cells, we constructed non-expressing retroviral vectors that have LacZ and neomycin resistance gene (neo) sequences but were modified to prevent protein expression [39]. In naive monkeys receiving autologous lymphocytes transduced with a neo-expressing vector and the non-expressing vector, lymphocytes transduced with the non-expressing vector persisted for more than 1 year whereas lymphocytes transduced with the neo-expressing vector disappeared or decreased to minimal levels within several weeks after infusion. The elimination of neo-transduced cells was shown to be mediated by cellular immune responses [40].

However, when we transduced HSCs with both the neo-expressing and non-expressing vectors followed by myeloablation and transplantation, peripheral blood cells transduced with the neo-expressing vector as well as the non-expressing vector persisted for more than 1 year, despite that neo was expressed out to 1 year after transplantation. The neo-expressing cells were not eliminated, implying that tolerance to the neo product was induced in the animals when HSCs were transduced with neo instead of lymphocytes.

To confirm that tolerance to the neo product was obtained in the animals, we proceeded to autologous lymphocyte transduction with another neo-expressing vector followed by reinfusion into the same monkeys to see how long transduced cells persisted in vivo. In animals first transplanted with transduced stem cells, neo-marked lymphocytes persisted for half a year to date, although lymphocytes transduced with an unrelated GFP-expressing vector disappeared several weeks after infusion. These results suggest that transfer of a xenogeneic gene such as neo introduced via HSCs induces specific tolerance to the gene product [39].

Therefore, when HSCs are used as a target of gene therapy, it is possible to perform replacement gene therapy to patients who have a null-type defect of a certain protein and thus do not have immunological tolerance to it.

The method of inducing specific tolerance by introduction of genes into HSCs may also pave the way for treatment of patients with some allergic and autoimmune diseases.

References

1. Dunbar CE, Emmons RVB (1994) Gene transfer into hematopoietic progenitor and stem cells. Stem Cells (Basel) 12:563–576
2. Naldini L, Bloemer U, Gallay P, Ory D, Mulligan R, Gage FH, Verma IM, Trono D (1996) In vivo gene delivery and stable transduction of nondividing cells by a lentiviral vector. Science 272:263–267
3. Williams DA, Lemischka IR, Nathan DG, Mulligan RC (1984) Introduction of new genetic materials into pluripotent hematopoietic stem cells of the mouse. Nature (Lond) 310:476–480
4. Dunbar CE, Cottler-Fox M, O'Shaughnessy JA, Doren S, Carter C, Berenson R, Brown S, Moen RC, Greenblatt J, Stewart FM, Leitman SF, Wilson WH, Cowan K, Young NS, Nienhuis AW (1995) Retrovirally marked CD34-enriched peripheral blood and bone marrow cells contribute to long-term engraftment after autologous transplantation. Blood 85:3048–3057
5. Abkowitz JL, Catlin SN, Guttorp P (1996) Evidence that hematopoiesis may be a stochastic process in vivo. Nat Med 2:190–197
6. Miller AD (1996) Cell surface receptors for retroviruses and implications for gene transfer. Proc Natl Acad Sci USA 93:11407–11413
7. Orlic D, Girard LJ, Jordan CT, Anderson SM, Cline AP, Bodine DM (1996) The level of mRNA encoding the amphotropic retrovirus receptor in mouse and human hematopoietic stem cells is low and correlates with the efficiency of retrovirus transduction. Proc Natl Acad Sci USA 93:11097–11102
8. Crooks GM, Kohn DB (1993) Growth factors increase amphotropic retrovirus binding to human CD34+ bone marrow progenitor cells. Blood 82:3290–3297
9. Shah AJ, Smogorzewska EM, Hannum C, Crooks GM (1996) Flt3 ligand induces proliferation of quiescent human bone marrow CD34+CD38– cells and maintains progenitor cells in vitro. Blood 87:3563–3570
10. Rusten LS, Lyman SD, Veiby OP, Jacobsen SE (1996) The FLT3 ligand is a direct and potent stimulator of the growth of primitive and committed human CD34+ bone marrow progenitor cells in vitro. Blood 87:1317–1325
11. Petzer AL, Zandstra PW, Piret JM, Eaves CJ (1996) Differential cytokine effects on primitive (CD34+CD38–) human hematopoietic cells: novel responses to Flt3-ligand and thrombopoietin. J Exp Med 183:2551–2558
12. Sitnicka E, Lin N, Priestley GV, Fox N, Broudy VC, Wolf NS, Kaushansky K (1996) The effect of thrombopoietin on the proliferation and differentiation of murine hematopoietic stem cells. Blood 87:4998–5005
13. Ramsfjell V, Borge OJ, Cui L, Jacobsen SE (1997) Thrombopoietin directly and potently stimulates multilineage growth and progenitor cell expansion from primitive (CD34+ CD38–) human bone marrow progenitor cells: distinct and key interactions with the ligands for c-kit and flt3, and inhibitory effects of TGF-beta and TNF-alpha. J Immunol 158:5169–5177
14. Hanenberg H, Xiao XL, Dilloo D, Hashino K, Kato I, Williams DA (1996) Colocalization of retrovirus and target cells on specific fibronectin fragments increases genetic transduction of mammalian cells. Nat Med 2:876–882

168   Y. Hanazono et al.

15. Yu J, Soma T, Hanazono Y, Dunbar CE (1998) Abrogation of TGF-β activity during
    retroviral transduction improves murine hematopoietic progenitor and repopulating
    cell gene transfer efficiency. Gene Ther 5:1265–1271
16. Dao MA, Nolta JA (1998) Reduction in levels of the CDK inhibitors p15^{INK4B} and
    p27^{KIP-1} allows transduction of primitive, reconstituting hematopoietic cells. Blood
    92(suppl 1):521a
17. Tisdale JF, Hanazono Y, Sellers SE, Agricola BA, Metzger ME, Donahue RE, Dunbar CE
    (1998) Ex vivo expansion of genetically marked rhesus peripheral blood progenitor
    cells results in diminished long-term repopulating ability. Blood 92:1131–1141
18. Kiem H-P, Andrews RG, Morris J, Peterson L, Heyward S, Allen JM, Rasko JEJ, Potter J,
    Miller AD (1998) Improved gene transfer into baboon marrow repopulating cells using
    recombinant human fibronectin fragment CH-296 in combination with interleukin-6,
    stem cell factor, FLT-3 ligand, and megakaryocyte growth and development factor.
    Blood 92:1878–1886
19. Sharma S, Cantwell M, Kipps TJ, Friedmann T (1996) Efficient infection of human T-
    cell line and of human primary peripheral blood leukocytes with a pseudotyped retro-
    virus vector. Proc Natl Acad Sci USA 93:11842–11847
20. Bunnell BA, Muul LM, Donahue RE, Blaese RM, Morgan RA (1995) High-efficiency
    retroviral-mediated gene transfer into human and nonhuman primate peripheral
    blood lymphocytes. Proc Natl Acad Sci USA 92:7739–7743
21. Kiem HP, Heyward S, Winkler A, Potter J, Allen JM, Miller AD, Andrews RG (1997) Gene
    transfer into marrow repopulating cells: comparison between amphotropic and gibbon
    ape leukemia virus pseudotyped retroviral vectors in a competitive repopulation assay
    in baboons. Blood 90:4638–4645
22. Crooks GM, Case SS, Price MA, Jordan CT, Bauer G, Xu D, Yu XJ, Verma IM, Naldini L,
    Kohn DB (1998) Transduction of CD34+CD38– cells with lentiviral vectors. Blood
    92(suppl 1):521a
23. Emerson SG (1996) Ex vivo expansion of hematopoietic precursors, progenitors, and
    stem cells: the next generation of cellular therapeutics. Blood 87:3082–3088
24. Holyoake TL, Alcorn MJ, Richmond L, Farrell E, Pearson C, Green R, Dunlop DJ,
    Fitzsimons E, Pragnell IB, Franklin IM (1997) CD34 positive PBPC expanded ex vivo
    may not provide durable engraftment following myeloablative chemoradiotherapy
    regimens. Bone Marrow Transplant 19:1095–1101
25. Migita M, Medin JA, Pawliuk R, Jacobson S, Nagle JW, Anderson S, Amiri M, Humphries
    RK, Karlsson S (1995) Selection of transduced CD34+ progenitors and enzymatic cor-
    rection of cells from Gaucher patients, with bicistronic vectors. Proc Natl Acad Sci USA
    92:12075–12079
26. Anderson SM, Yu G, Giattina M, Miller JL (1996) Intercellular transfer of a glyco-
    sylphosphatidylinositol (GPI) -linked protein: release and uptake of CD4-GPI from
    recombinant adeno-associated virus-transduced HeLa cells. Proc Natl Acad Sci USA
    93:5894–5898
27. Chalfie M, Tu Y, Euskirchen G, Ward WW, Prasher DC (1994) Green fluorescent protein
    as a marker for gene expression. Science 263:802–805
28. Hanazono Y, Dunbar CE, Emmons RVB (1997) Green fluorescent protein retroviral
    vectors: low titer and high recombination frequency suggest a selective disadvantage.
    Hum Gene Ther 8:1313–1319
29. Kume A, Xu R, Matsuda K, Ueda Y, Urabe M, Suda T, Ozawa K (1998) In vivo tracking
    of retrovirally transduced hematopoietic cells stably expressing CD24 and GFP trans-
    genes (abstract). In: 1st annual meeting of the American Society of Gene Therapy,
    Seattle, WA, USA, 90a

30. Persons DA, Allay JA, Riberdy JM, Wersto RP, Donahue RE, Sorrentino BP, Nienhuis AW (1998) Use of the green fluorescent protein as a marker to identify and track genetically modified hematopoietic cells. Nat Med 4:1201–1205

31. Donahue RE, Wersto RP, Allay JA, Agricola BA, Metzger ME, Persons DA, Nienhuis AW, Sorrentino BP (1998) High proportions of circulating leukocytes express a green fluorescent protein-containing retroviral vector in non-human primates transplanted with transduced peripheral blood CD34+ cells. Blood 92(suppl 1):690a

32. Phillips K, Gentry T, McCowage G, Gilboa E, Smith C (1996) Cell-surface markers for assessing gene transfer into human hematopoietic cells. Nat Med 2:1154–1156

33. Bonini C, Ferrari G, Verzeletti S, Servida P, Zappone E, Ruggieri L, Ponzoni M, Rossini S, Mavilio F, Traversari C, Bordignon C (1997) HSV-TK gene transfer into donor lymphocytes for control of allogeneic graft-versus-leukemia. Science 276:1719–1724

34. Sorrentino BP, Brandt SJ, Bodine D, Gottesman M, Pastan I, Cline A, Nienhuis AW (1992) Selection of drug-resistant bone marrow cells in vivo after retroviral transfer of human MDR1. Science 257:99–103

35. Bunting KD, Galipeau J, Topham D, Benaim E, Sorrentino BP (1998) Transduction of murine bone marrow cells with an MDR1 vector enables ex vivo stem cell expansion, but these expanded grafts cause a myeloproliferative syndrome in transplanted mice. Blood 92:2269–2279

36. Allay JA, Persons DA, Galipeau J, Riberdy JM, Ashmun RA, Blakley RL, Sorrentino BP (1998) In vivo selection of retrovirally transduced hematopoietic stem cells. Nat Med 4:1136–1143

37. Ito K, Ueda Y, Kokubun M, Urabe M, Inaba T, Mano H, Hamada H, Kitamura T, Mizoguchi H, Sakata T, Hasegawa M, Ozawa K (1997) Development of a novel selective amplifier gene for controllable expansion of transduced hematopoietic cells. Blood 90:3884–3892

38. Riddell SR, Elliott M, Lewinsohn DA, Gilbert MJ, Wilson L, Manley SA, Lupton SD, Overell RW, Reynolds TC, Corey L, Greenberg PD (1996) T-cell mediated rejection of gene-modified HIV-specific cytotoxic T lymphocytes in HIV-infected patients. Nat Med 2:216–223

39. Hanazono Y, Heim D, Tisdale J, Giri N, Sellers S, Agricola B, Metzger M, Donahue R, Dunbar C (1998) Equivalent in vivo persistence of progeny of CD34+ cells transduced with expressed versus non-expressed xenogeneic genes, with induction of tolerance to subsequent xenogeneic-gene expressing lymphocytes. Blood 92(suppl 1):148a

40. Heim DA, Hanazono Y, Childs R, Metzger M, Donahue RE, Dunbar CE (1998) In vivo persistence of rhesus monkey lymphocytes transduced with a non-expressing retroviral vector compared to rapid clearance of lymphocytes transduced with a neo-expressing vector. Blood 92(suppl 1):688a

Ex Vivo Expansion of Hematopoietic Cells

Ian McNiece, Roy Jones, Pablo Cagnoni, Scott Bearman, Yago Nieto, and Elizabeth J. Shpall

Summary. Ex vivo expanded progenitor cells have been proposed as a source of cells to support high-dose chemotherapy and to decrease or eliminate the period of neutropenia following transplantation. To date, no clinical studies using ex vivo expanded cells have demonstrated any decrease in the time to neutrophil or platelet recovery, although a number of clinical studies have been performed using a variety of growth factor cocktails and culture conditions. During the past 6 years we have developed a static culture system that results in optimal expansion of myeloid progenitor cells as measured by GM-CFC in vitro. We have initiated two clinical studies to evaluate this culture system in cancer patients receiving autologous peripheral blood progenitor cells (PBPC) or allogeneic cord blood cells (CB) to support high-dose chemotherapy. CD34 selected cells were cultured for 10 days in 800 ml of defined media containing 100 ng/ml each of rhSCF, rhG-SCF, and rhMGDF in 1-l teflon bags. After culture the cells were washed with 3 volumes of PBS to remove all media and growth factors and reinfused with daily administration of rhG-CSF. In both studies, unexpanded cells were given in addition to the expanded cells to ensure durability of the graft. Patients transplanted with expanded PBPC cells recovered neutrophil counts on day 5 (3 patients) or day 6 (3 patients) post transplant. The median time to neutrophil engraftment in historical controls was 11 days, with the earliest recovery at day 7. No effect on platelet recovery has been observed in any patients to date. These data demonstrate that PBPC expanded under the conditions defined can significantly shorten the time to engraftment of neutrophils. In the second study adult (weight, 46–116 kg) patients were transplanted with expanded CB cells (40% of product), while 60% of the product was transplanted unmanipulated.

BMT Program, University of Colorado Health Sciences Center, 4200 East Ninth Avenue, Denver, CO 80262, USA

The following table summarizes the expansion and engraftment data for these patients:

Patient	Weight (kg)	Infused cells (×10⁷/kg)		Expansion of CFC (fold)	Engraftment	
		Day 0	Day 10		ANC	Platelets
1	116	0.35	0.09	26	34	62+
2	74	0.4	0.29	10	21	51
3	86	0.37	0.31	40	26	91
4	59	0.78	0.44	31	21	23+
5	60	0.70	0.40	22	23	102+
6	11	4.9	3.7#	7	15	56+

Reinfusion of the expanded cells was not associated with any adverse events. Neutrophil engraftment (days to an absolute neutrophil count [ANC] >500/µl) occurred between days 21 to 34 in patients 1 to 5 and even earlier in patient 6, who recovered to an ANC >500 on day 15. This rapid engraftment in patient 6 is consistent with the infusion of expanded cells on day 0 compared to day +10 in the other patients. Previous studies by Gluckman and colleagues reported that for patients of more than 45 kg in weight, only 11 of 23 and 5 of 23 achieved neutrophil and platelet engraftment, respectively, by day 60; this suggests that expanded CB cells may provide more rapid engraftment in larger patients, with 5 of 5 achieving neutrophil engraftment by day 34 or earlier. The availability of CB products frozen in aliquots may enable more rapid engraftment, similar to the recovery of patient 6. Other studies are in progress that suggest a role of ex vivo expanded cells in cellular support for high-dose chemotherapy. The optimal use of expanded cells, however, still remains to be defined.

Key words. Ex vivo expansion, Cord blood, PBPC, Clinical, Transplantation

Introduction

The marrow is the principal site for blood cell formation in humans. In normal adults the body produces about 2.5 billion red blood cells (RBC), 2.5 billion platelets (plts), and 10 billion granulocytes per kilogram of body weight per day [1]. The production of mature blood cells is a continual process that is the result of proliferation and differentiation of stem cells, committed progenitor cells, and differentiated cells (Fig. 1). Within these three stages extensive expansion of cell numbers occurs through cell division. A single stem cell has been proposed to be capable of 50 cell divisions or doublings and has the capacity to generate up to 10^{15} cells, or sufficient cells for as long as 27 years [2]. The proliferation and differentiation of cells is controlled by a group of proteins called hematopoietic growth factors (HGFs). Many of the

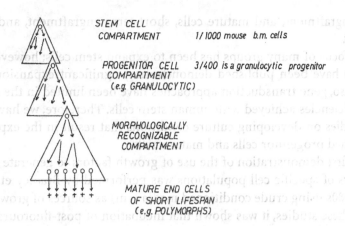

FIG. 1. Model of expansion within hematopoietic compartments

HGFs have been isolated and cloned and are large amounts of recombinant proteins are available for clinical use or ex vivo manipulations of cells. If we could replicate this cell amplification in vitro with the HGFs, it may be possible to generate large numbers of cells that could be used for a variety of clinical applications. These include the following:

Supplementing stem cell grafts with more mature precursors to shorten or
    potentially prevent chemotherapy-induced pancytopenia
Increasing the number of primitive progenitors to ensure hematopoietic
    support for multiple cycles of high-dose therapy
Obtaining a sufficient number of stem cells from a single marrow aspirate or
    pheresis procedure, thus reducing the need for large-scale harvesting of
    marrow or multiple leukapheresis
Generating sufficient cells from a single cord blood unit to reconstitute an
    adult following high-dose chemotherapy
Purging stem cell products of contaminating tumor cells
Generating large volumes of immunologically active cells with antitumor
    activity to be used in immunotherapeutic regimens
Increasing the pool of stem cells that could be targets for the delivery of gene
    therapy

It is important, therefore, to consider the clinical application when discussing ex vivo expansion approachs and to define culture conditions that are relevant to the particular cell type being expanded. The relevant properties of the cells in each of the different hematopoietic compartments are summarized as follows: stem cells, long-term engraftment and targets for gene replacement therapy; committed progenitor cells, intermediate and short-term engraftment; and mature cells, short-term engraftment, and immune therapy.

The focus of many groups has been to expand stem cells; however, to date no data have been published demonstrating significant expansion of stem cells. Also, gene transduction approaches have been limited in the transduction efficiencies achieved with human stem cells. Therefore, we have focused our studies on developing culture conditions that result in the expansion of committed progenitor cells and mature cells.

The first demonstration of the use of growth factors to generate increased numbers of specific cell populations was performed by Bradley et al. in the early 1980s using crude conditioned media (cm) as sources of growth factors [3]. In these studies, it was shown that incubation of post-fluorouracil (post-FU) mouse bone marrow cells in WEHI-3 CM for 7 days resulted in a 60-fold increase of CFU-S$_{13}$ and a 53-fold increase in GM-CFC. In subsequent studies from this group, it was shown that preincubation with HGFs (also crude CM) could expand primitive murine progenitor cells called high proliferative potential colony-forming cells (HPP-CFC) and cells with in vivo marrow repopulating ability [4,5]. Using a similar culture system of human bone marrow cells in Teflon bottles, it was shown in 1988 that the combination of rhGM-CSF plus rhIL-3 could generate a 7-fold increase in committed progenitor cells (granulocyte-macrophage colony-forming cells, GM-CFC) [6]. In 1991, Bernstein et al. [7] showed that incubation of single CD34+Lin− cells in the combination of interleukin-3 (IL-3), granulocyte colony-stimulating factor (G-CSF), and stem cell factor (SCF) gave rise to an increase of 10 fold of colonies in vitro.

The use of ex vivo expansion to generate mature neutrophil precursors was proposed in 1992 by Haylock et al. [8]. These authors demonstrated that the combination of IL-1, IL-3, IL-6, GM-CSF, and SCF could generate a 1 324-fold increase in nucleated cells and a 66-fold increase in GM-CFC. The cells produced under these conditions were predominantly neutrophil precursors. The culture conditions used were static cultures and utilized CD34+ cells as the starting population. Several investigators have demonstrated the requirement for CD34 selection of the starting cells for optimal expansion [8–11]. Subsequent studies were performed at a clinical scale using optimal culture conditions in Teflon bags and with fully defined media appropriate for clinical applications [12]. This work utilized the growth factor cocktail composed of SCF, G-CSF, and megakavyocyte growth and development factor (MGDF) [12]. Other cocktails of growth factors are effective in expanding CD34+ cells; however, the availability of clinical-grade growth factors has been limited by commercial considerations.

The in vivo potential of ex vivo expanded cells was first reported in murine studies by Muench et al. [13]. This study demonstrated that bone marrow cells

TABLE 1. Engraftment of irradiated baboons with ex vivo expanded peripheral blood progenitor cells (PBPC)

Group (n = 3)	In vitro expansion	Posttransplant G-CSF + MGDF	Days ANC <500/µl	ANC ≥1 000/µl	Platelets ≥1 000/µl	Nadir ANC (/µl)
I	No	No	7, 16, 16	15, 28, 26	14, 8, 26	9 ± 16
II	Yes	No	13, 17, 12	24, 70, 21	NR, 42, 34	72 ± 47
III	No	Yes	6, 8, 16	10, 13, 22	9, 17, NR	50 ± 40
IV	Yes	Yes	8, 0, 0	17, 9, 8	30, 25, 22	461 ± 154

expanded in SCF plus IL-1 engrafted lethally irradiated mice and were capable of sustaining hematopoiesis over the long term in these animals. In addition, the bone marrow from these engrafted mice could repopulate secondary recipients. The authors concluded that the expansion of mouse bone marrow cells did not adversely effect the proliferative capacity and lineage potential of the stem cell compartment [13].

Recent studies in normal baboons [14] have demonstrated the potential clinical benefit of ex vivo expanded cells. Andrews and colleagues harvested PBPC from G-CSF-mobilized normal baboons and expanded the CD34+ cells for 10 days in SCF plus G-CSF plus MGDF. After the culture period, the cells were washed and infused into the baboons after lethal irradiation. The fold-expansion obtained was low compared to human CD34+ cells and probably as the result of species variations of the growth factors and cell behavior in culture conditions developed for expansion of human cells; GM-CFC were expanded seven- to eightfold. Table 1 summarizes the engraftment characteristics of the different treatment groups in these studies.

Group IV, transplanted with expanded CD34+ cells and given posttransplant G-CSF and MGDF, had a significantly shorter duration of neutropenia and significantly higher WBC and polymorphonuclear neutrophil (PMN) nadirs compared to animals in the other groups; in fact, two of the three animals had no days with neutrophils below 500/µl, a clinical endpoint used for neutrophil engraftment. In these studies, in vitro expansion did not influence platelet recovery despite the use of MGDF in both cultures and after transplantation. Further studies are needed to determine the culture conditions that enhance platelet recovery of PBPC.

Transplantation of Ex Vivo Expanded PBPC in Patients Receiving High-Dose Chemotherapy

Two separate studies have recently been reported at the American Society of Hematology describing the expansion of PBPC from patients undergoing transplantation for high dose chemotherapy (HDC) [15,16]. Both these studies used identical culture conditions, namely culture of CD34+ cells in defined media (DM) (Amgen, Thousand Oaks, CA) supplemented with the growth factors rhSCF, rhG-CSF, and rhMGDF at 100 ng/ml for 10 days in Teflon bags (American Fluoroceal, Columbia, MD, USA).

University of Colorado

Patients with high risk stage II, III, or IV breast cancer who were appropriate candidates for high-dose chemotherapy requiring cellular support were eligible. All patients had received prior adjuvant chemotherapy. Table 2 summaries the patients' disease, mobilization regimen, and chemotherapy regimen, All patients were mobilized with rhG-CSF (Filgrastim, Amgen; 10 μg/kg/day) alone or in combination with rhSCF (Stemgen, Amgen; 25 μg/kg/day) for 9 days and underwent leukapheresis on days 5 through 9. Following mobilization all patients received high-dose chemotherapy consisting of the following: (i) phase II patients with stage IV NED (no evidence of disease) ($n = 1$) or stage II–III breast cancer involving ≥10 ($n = 5$) or 4–9 ($n = 2$) axillary lymph nodes; cyclophosphamide ($1875 \, \text{mg/m}^2/\text{day}$ IV ×3, administered on days −6, −5, −4), cisplatin ($55 \, \text{mg/m}^2/\text{day}$ IV ×3, administered on days −6, −5, −4), and BCNU (carmustine) ($600 \, \text{mg/m}^2$ IV ×1, given on day −3), or (ii) phase I patients with stage IV disease refractory to induction therapy ($n = 1$) or more than 1 prior regimen for metastases ($n = 1$); taxotere (200–$300 \, \text{mg/m}^2$ on day −7), carboplatin ($1\,000 \, \text{mg/m}^2/\text{day}$ as CI on days −6, −5, −4), and melphalan ($150 \, \text{mg/m}^2/\text{day}$ on days −6, −5, −4), or (iii) phase II patients with inflammatory breast cancer ($n = 1$) or stage IV disease responding to induction therapy ($n = 1$); taxol ($725 \, \text{mg/m}^2/\text{day}$ on day −7), cyclophosphamide ($1\,875 \, \text{mg/m}^2/\text{day}$, on days −6, −5, −4), and cisplatin ($55 \, \text{mg/m}^2/\text{day}$, on days −6, −5, −4).

Leukapheresis products (LP) were harvested on days 5 through 9, with CD34+ cell selection performed on the first four LP. The fifth LP was frozen

TABLE 2. Breast cancer patients ($n = 12$): characteristics and prior therapy

Stage	n	No. prior cycles medication
II/III:		
4–9 node+	2	
≥10 node+	6	
Inflammmatory	1	
Total	9	4
Ivned	1	7
IV measurable	2[a]	16, 12

NED, no evidence of disease.
[a] Enrolled on phase I chemotherapy trials; all other patients on phase II studies.

unselected as a backup product. CD34+ selection was performed using the Isolex 300i (Nexell Irvine, CA, USA). After selection, each product was frozen in liquid nitrogen. On day −10 of treatment, two LP were thawed and placed into ex vivo expansion culture. The cells were diluted in defined media (DM; Amgen) supplemented with 100 ng/ml each of rhSCF, rhMGDF, and rhG-CSF to 20 000 cells/ml in 800 ml of media and transferred into Teflon bags (American Fluoroceal). The bags were incubated at 37°C for 10 days in a 5% CO_2 incubator. On day 0 of treatment, the cultures were harvested using a cell washer (Cobe) and the media and growth factors removed with washing.

Following ex vivo expansion of the CD34 selected cells, patients in cohort 1 ($n = 10$) were reinfused with expanded cells on day 0 followed by the unexpanded CD34+ cells (Fig. 2a). Patients in cohort 2 ($n = 10$) received only ex vivo expanded cells, and the unexpanded CD34+ cells were maintained frozen in liquid nitrogen as a backup source of hematopoietic cells (Fig. 2b).

Transplantation of ex vivo expanded PBPC resulted in rapid engraftment of neutrophils (ANC >500/μl) with one patient engrafting on day 4 and a

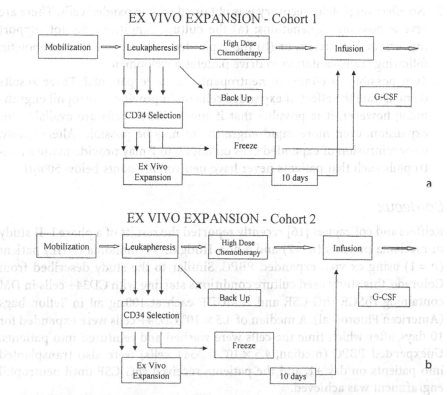

FIG. 2a,b. Cohorts for ex vivo expansion of peripheral blood progenitor cells (PBPC)

number of patients engrafting on days 5 and 6. The median time of neutrophil engraftment was day 6 in the first 15 patients. Our historical controls have a median time to neutrophil engraftment of 9 days ($n = 175$), with a range of 7 to 24 days. The patients in cohort 2 were transplanted with only expanded cells, and the first couple of patients are now at more than 70 days post transplant with each patient maintaining a durable graft. These patients will be monitored over the long term to determine if the expansion of the cells compromises the long-term engrafting cells in these products. No effect on platelet recovery has been observed. The effect of tumor cell depletion was evaluated in all patients; however, no PBPC products contained any detectable tumor cells before selection, following selection, or following expansion.

The results of this study raise several questions that remain to be answered.

1. The effect of the expansion culture on long-term engrafting cells. The fate of stem cells during the expansion cultures has not been determined; however, cohort 2 patients are only receiving expanded cells and will provide definitive data on the long-term durability of the expanded products.
2. No effect on platelet recovery was obtained with expanded cells. There are several possible explanations: (a) the culture conditions do not support precursors of platelets, or (b) the expanded cells require thrombopoietin following transplantation to drive platelet development.
3. Is it possible to eliminate neutropenia in these patients? These results demonstrate the effect of expanded cells on improving neutrophil engraftment; however, it is possible that if more CD34+ cells are available for expansion, even more rapid engraftment may be possible. Alternatively, daily reinfusion of expanded cells on days 0 to 7 may provide mature neutrophils such that patients never have neutrophil counts below 500/μl.

Bordeaux

Reiffers and colleagues [16] recently reported the results of a phase I–II study in myeloma patients ($n = 7$) and a non-Hodgkin's lymphoma (NHL) patient ($n = 1$) using ex vivo expanded PBPC. Similar to the study described from Colorado, this study used culture conditions starting with CD34+ cells in DM containing rhSCF, rhG-CSF, and rhMGDF each at 100 ng/ml in Teflon bags (American Fluoroceal). A median of 4.5×10^6 CD34+ cells were expanded for 10 days, after which time the cells were washed and reinfused into patients. Unexpanded PBPC (median, 4.5×10^6 CD34+ cells) were also transplanted into patients on day +1, and the patients received rhG-CSF until neutrophil engraftment was achieved.

The posttransplant neutropenia was markedly reduced in these patients compared to historical controls. The median number of days of neutropenia (ANC, 500/µl) was 2 days with a range of 0 to 5 days. Three of the eight patients never had a neutrophil count below 500. The median time to neutrophil recovery (ANC > 500/µl) was 5 days. No effect on platelet recovery was seen in these patients, consistent with the data described previously [15]. This study continues to accrue patients.

Transplantation of Ex Vivo Expanded Cord Blood Cells in Cancer Patients

Cancer patients chronic myeloid leukemia [CML], $n = 2$; chronic lymphocytic leukemia [CLL], $n = 1$; NHL, $n = 3$; acute lymphocytic leukemia [ALL], $n = 2$; breast cancer, $n = 2$) who were appropriate candidates for high-dose chemotherapy requiring cellular support were eligible [17]. The preparative regimens used were (i) total body irradiation 1 200 cGy, melphelan 140 mg/m², ATG 90 mg/m² ($n = 7$); or (ii) Busulfan 16 mg/kg, melphelan 140 mg/m², ATG 90 mg/m² ($n = 3$).

The cord blood products were obtained from several cord blood banks including the New York Blood Center ($n = 5$), Dusseldorf Cord Blood Bank ($n = 2$), the Australian Cord Blood Bank ($n = 1$), and the University of Colorado Cord Blood Bank ($n = 2$). The cord products were frozen as a single product ($n = 6$) or in aliquots of $2 \times 50\%$ ($n = 2$) or in 40% and 60% aliquots ($n = 2$). The CB products frozen in a single product were thawed on day 0 and 60% reinfused unmanipulated, while the remaining 40% was CD34 selected and placed into ex vivo expansion culture (Fig. 3a). The other CB products were thawed at different times, with a 50% or 40% aliquot thawed on day -10 for expansion culture and the remaining aliquot thawed on day 0 and reinfused without manipulation (Fig. 3b). All fractions for expansion were CD34 selected using the Nexell Isolex 300i, and the CD34 selected cells were placed into a single Teflon bag containing 800 ml of defined media (Amgen) containing 100 ng/ml each of rhSCF, rhG-CSF, and rhMGDF. The bags were incubated for 10 days in a 5% CO_2 incubator, after which time the cells were harvested, washed, and reinfused. Following transplantation all patients received GVHD prophylaxis consisting of (i) cyclosporine 5 mg/kg IV q12h starting day 2 and high-dose steroids ($n = 8$), or (ii) cyclosporine 5 mg/kg IV q12h starting day 2 and moderate-dose steroids ($n = 2$).

The patients received a median total cell number of 0.94×10^7 cells/kg (range, 0.43 to 8.4×10^7/kg total cells). All patients have achieved neutrophil engraftment with a median time of 24 days (range, 15–34 days) and to platelet engraftment of 82 days. Considering the low numbers of cells infused per

EX VIVO EXPANSION -Strata A

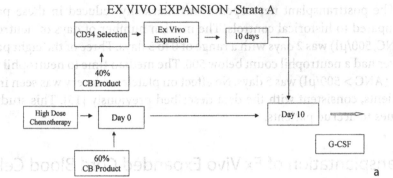

EX VIVO EXPANSION -Strata B

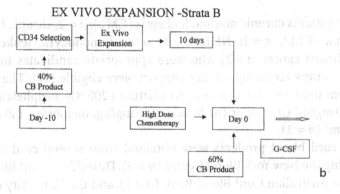

FIG. 3a,b. Cohorts for ex vivo expansion of cord blood cells (CB)

kilogram in these patients, the time to neutrophil engraftment was faster than previous reports in patients with weights greater than 45 kg [18]. In particular, we are following the recovery of patients in cohort 2 who received the expanded CB cells on day 0 as the first patient in this cohort engrafted neutrophils on day 15. Of the ten patients treated, four patients are alive without disease ranging from 70 to 430 days, four patients are too early to evaluate, and two patients died of causes unrelated to the expanded cells.

Ex Vivo Expansion in Perfusion Culture: The Aastrom Expansion System

Several abstracts were reported at the American Society of Hematology (ASH) meeting in Miami describing studies of ex vivo expansion using the Aastrom Cell Production System for CB and PBPC. Kurtzberg and colleagues [19] described the expansion of CB products that were transplanted into 21 patients. Each patient received unmanipulated CB cells on days 0 and the expanded cells on day +12. No significant effects on engraftment kinetics were observed in these patients. Stiff and colleagues [20] also reported on expansion of CB cells in the Aastrom system for nine patients. The median weight of the patients was 74 kg (range, 47–117 kg); the median time to neutrophil engraftment was 26 days (range, 14–36 days). Engraftment of platelets was delayed in these patients. The authors concluded that ex vivo expanded CB cells may be useful in adults with otherwise incurable hematological disorders.

The Aastrom system has also been used for the expansion of bone marrow. Studies have demonstrated that expansion of a small aliquot of bone marrow (BM) cells can provide short- and long-term engraftment following myeloablative chemotherapy [21]. Other studies have combined the expanded BM cells with single PBPC products, reporting that the patients engrafted neutrophils and platelets equivalent to optimal PBPC products. The clinical utility of this approach remains to be fully defined.

Conclusion

The studies outlined here demonstrate the potential clinical utility of ex vivo expanded cells. The optimal growth factor cocktail for expansion of different cell types is still to be determined. Several studies show a very significant effect on neutrophil engraftment; however, it is still unclear what cells are responsible for this more rapid recovery. Studies are needed to determine how to evaluate expanded products to define the cell type that leads to more rapid neutrophil engraftment. Also, culture conditions still must be identified that will allow for the expansion of a megakaryocyte/platelet precursor which will provide more rapid platelet engraftment. In summary, there are exciting clinical results that suggest expanded cells have a major application in cellular support for high-dose chemotherapy, and further clinical studies are needed to better define what cells are required and what culture conditions enable suitable expansion of these cell types.

References

1. Williams WJ, Beutler E, Erslev AJ, Lichtman MA (eds) (1990) Hematology, 4th edn. McGraw-Hill, New York
2. Kay HEM (1965) How many cell generations? Lancet 2:418–421
3. Bradley TR, Hodgson GS, Kriegler AB, McNiece IK (1985) Generation of CFU-S$_{13}$ in vitro. In: Hematopoietic stem cell physiology. Liss, New York, pp 39–56
4. McNiece IK, Bradley TR, Kriegler AB, Hodgson GS (1986) Subpopulations of mouse bone marrow high-proliferative-potential colony-forming cells. Exp Hematol 14:856
5. McNiece IK, Williams NT, Johnson GR, Kriegler AB, Bradley TR, Hodgson GS (1987) Generation of murine hematopoietic precursor cells from macrophage high-proliferative-potential colony-forming cells. Exp Hematol 15:972

6. McNiece IK, Andrews RG, Stewart FM, Quesenberry PQ (1987) Synergistic interactions of human growth factors in *in vitro* cultures of human bone marrow cells. Blood 72(5) (suppl 1):125a

7. Bernstein ID, Andrews RG, Zsebo KM (1991) Recombinant human stem cell factor enhances the formation of colonies by CD34+ and CD34+lin– cells, and the generation of colony-forming cell progeny from CD34+lin– cells cultured with interleukin-3, granulocyte colony-stimulating factor, or granulocyte-macrophage colony-stimulating factor. Blood 77:2316

8. Haylock DN, To LB, Dowse TL, Juttner CA, Simmons PJ (1992) Ex vivo expansion and maturation of peripheral blood CD34+ cells into the myeloid lineage. Blood 80(6):1405

9. Purdy MH, Hogan CJ, Hami L, McNiece I, Franklin W, Jones W, Bearman S, Berenson RJ, Cagnoni PJ, Heimfeld S, Shpall EJ (1995) Large scale ex vivo expansion of CD34-positive hematopoietic progenitor cells for transplantation. J Hematother 4:515–525

10. Briddell RA, Kern BP, Zilm KL, Stoney GB, McNiece IK (1996) Purification of CD34+ cells is essential for optimal ex vivo expansion of umbilical cord blood cells. Exp Hematol 24(9):1055

11. Shieh J-H, Chen Y-F, Briddell R, Stoney G, McNiece I (1994) High purity of blast cells in CD34 selected populations are essential for optimal ex vivo expansion of human GM-CFC. Exp Hematol 22(8):756a

12. Stoney GB, Briddell RA, Kern BP, Zilm KL, McNiece IK (1996) Clinical scale ex vivo expansion of myeloid progenitor cells and megakaryocytes under GMP conditions. Exp Hematol 24(9):1043a

13. Muench MO, Firpo MT, Moore MAS (1993) Bone marrow transplantation with interleukin-1 plus kit-ligand ex vivo expanded bone marrow accelerates hematopoietic reconstitution in mice without the loss of stem cell lineage and proliferative potential. Blood 81:3463

14. Andrews RG, Briddell RA, Gough M, McNiece IK (1997) Expansion of G-CSF mobilized CD34+ peripheral blood cells (PBC) for 10 days in G-CSF, MGDF and SCF prior to transplantation decreased post-transplant neutropenia in baboons. Blood 90:10(suppl 1):92a

15. McNiece I, Hami L, Jones R, Bearman S, Cagnoni P, Nieto Y, Shpall EJ (1998) Transplantation of ex vivo expanded PBPC after high dose chemotherapy results in decreased neutropenia. Blood 92(10)(suppl 1):126a

16. Reiffers J, Cailliot C, Dazey B, Duchez I, Pigneux A, Cousin T, Bussiere B, Boiron JM (1998) Infusion of expanded CD34+ selected cells can abrogate post myeloablative chemotherapy neutropenia in patients with hematologic malignancies. Blood 92(10)(suppl 1):126a

17. Shpall EJ, Quinones R, Hami L, Jones R, Bearman S, Cagnoni P, Giller R, Nieto Y, Roman-Unfer S, Ross M, McNiece I (1998) Transplantation of cancer patients receiving high dose chemotherapy with ex vivo expanded cord blood cells. Blood 92(10)(suppl 1):646a

18. Gluckman E, Rocha V, Boyer-Chammard A, Locatelli F, Arcese W, Pasquini R, et al (1997) Outcome of cord-blood transplantation from related and unrelated donors. N Engl J Med 337(6):373

19. Jaroscak J, Martin PL, Waters-Pick B, Armstrong RD, et al (1998) A phase I trial of augmentation of unrelated umbilical cord blood transplantation with ex-vivo expanded cells. Blood 92(10)(suppl 1):646a

20. Stiff P, Pecora A, Parthasarathy M, Preti R, Chen B, et al (1998) Umbilical cord blood transplants in adults using a combination of unexpanded and ex vivo expanded cells: preliminary clinical observations. Blood 92(10)(suppl 1):646a

21. Engelhardt M, Douville J, deReys S, von Kalle C, Behringer D, et al (1998) Transplantation of ex vivo perfusion culture expanded bone marrow cells produces durable hematopoietic reconstitution after myeloablative chemotherapy. Blood 92(10)(suppl 1):126a

Harnessing Graft-Versus-Leukemia Without Myeloablative Therapy

SERGIO A. GIRALT

Summary. High-dose chemotherapy and allogeneic progenitor cell transplantation has been used as a curative therapy for many patients with hematologic malignancies. However, because of the toxicity of myeloablative conditioning it had been limited to young patients in good medical condition. The use of less intensive nonablative regimens may provide a strategy that allows exploitation of the graft-versus-leukemia effect in older and debilitated patients.

Key words. Hematological malignancies, Malignancies, Nonmyeloablative conditioning, Bone marrow transplantation, Engraftment, Graft-vs-leukemia induction

Introduction

High-dose chemoradiotherapy with allogeneic bone marrow transplantation (BMT) is becoming increasingly more used as therapy for patients with hematological malignancies [1,2]. Although at first thought to be effective by virtue of the antileukemic effects of the dose-intensive myeloablative preparative regimens, it has been recognized that the curative potential of allogeneic BMT is mediated also by an immune-mediated graft-versus-leukemia effect. Evidence of a graft-versus-leukemia effect is demonstrated by the lower relapse rate in patients with graft-versus-host disease (GVHD) and the higher risk of relapse after syngeneic or T-cell-depleted transplants [3–8]. The most direct evidence for this effect is the reinduction of remission obtained with infusions of donor lymphocytes in patients relapsing after an allogeneic progenitor cell transplant. This effect is more pronounced in patients with chronic

Department of Blood and Bone Marrow Transplantation, University of Texas, M.D. Anderson Cancer Center, 1515 Holcombe Blvd., Houston, TX 77030, USA

myelogenous leukemia (CML), but has also been observed in patients with other hematological malignancies [9–17]. Thus, induction of a graft-versus-leukemia is sufficient to obtain long-term disease control without intense myeloablative conditioning, and could be explored in patients with suscepti-ble malignancies considered poor candidates for myeloablative therapies because of their age or concurrent medical conditions.

To exploit graft-versus-leukemia without myeloablative regimens, it would seem that engraftment of the allogeneic progenitor cells would be an impor-tant prerequisite. The minimum immunosuppressive dose of the most com-monly used agents necessary to allow engraftment of allogeneic progenitor cells in humans has not been well defined [18]. This dose probably depends on the degree of HLA compatibility between host and recipient, as well as progenitor cell dose, T-cell content of the graft, and probably unique patient characteristics determined by both the primary diagnosis and the prior therapy to which an individual patient has been exposed. Patients with severe aplastic anemia and Fanconi's syndrome have successfully engrafted with less intensive preparative regimens; however, graft failures and death from prolonged cytopenias occur in less than ideally matched situations. The main agents used in these situations have been cyclophosphamide in combi-nation with antithymocyte globulin with or without low doses of total body irradiation [19–21].

In the canine model, Storb and colleagues have pioneered the development of less toxic nonmyeloablative regimens using low doses of total body irradi-ation (TBI) in combination with posttransplant immunosuppression includ-ing cyclosporine and mycophenolate mofetil; in this model a minimum immunosuppressive dose of 200-cGy TBI was identified in the DLA-identical setting only if accompanied by posttransplant immunosuppressive therapies [22,23]. Other clinical observations suggesting a threshold immunosuppres-sive dose are fact that patients with aplastic anemia receiving HLA-identical sibling transplants have a very low rate of graft rejection with a preparative regimen consisting of cyclophosphamide (200 mg/kg) and antithrombocyte-globulm (ATG); however, this same preparative regimen is not immunosup-pressive enough to allow for engraftment of unrelated donor cells [20,24]. Thus, development of immunosuppressive nonmyeloablative regimens will require the combination of potent immunosuppressants, preferably with anti-tumor activity and other drugs at nonmyeloablative doses.

The purine analogs fludarabine and 2-chloroadenosine (2-CDA) would seem ideal candidates to explore in the development of nonmyeloablative immunosuppressant regimens that could facilitate engraftment while exert-ing their antitumor effects. Fludarabine is an effective agent for combination chemotherapy of AML and a potent immunosuppressive agent; studies at M.D. Anderson have established its role in treatment of a range of hematological malignancies [25–30]. 2-CDA has established activity in hairy cell leukemia, chronic lymphocytic leukemic (CLL), and other lymphoid malignancies [31,32]. Both purine analogs have been demonstrated to be potent immunosuppressants, producing profound lymphopenia, and inhibiting the mixed lymphocyte reaction in vitro [33,34]. We therefore postulated that nonablative chemotherapy using purine analog combinations would be sufficiently immunosuppressive to allow engraftment of allogeneic blood progenitor cells, and would be tolerable by older and medically debilitated patients ineligible for conventional myeloablative conditioning.

Nonablative Therapies for Myeloid Leukemias: M.D. Anderson Cancer Center Experience

At M.D. Anderson, we performed a pilot trial of purine analog containing nonmyeloablative therapy for patients with AML or CML considered ineligible for myeloablative therapy because of either age or medical condition [35]. Twenty-six patients have been treated, and their characteristics are summarized in Table 1.

TABLE 1. Characteristics of patients undergoing minitransplant for acute myeloid leukemia (AML) and chronic myelogenous leukemia (CML) at M.D. Anderson Cancer Center

n	26
Median age in years (range)	60 (28–72)
Diagnosis and stage at BMT	
AML/MDS	17
First relapse untreated	2
Untreated	1
Refractory relapse or >CR 2	14
CML	
First Chronic Phase	5
Transformed	4
Median time to transplant	490 days (77–3429)
Median no. prior therapies	2 (1–3)
Preparative regimen	
FlagIda	17
2CDA/AraC	9
Donor type and cell source	
Sib Full Match/1 Ag Mismatch	20/3
6/6 MUD	2
Syngeneic	1
PBSC/BM	22/4
GVHD Prophylaxis (twin excluded)	
CSA or CSA/MP	19
FK/MTX	6

Four of the 26 patients died before day 100, 3 of infectious complications and 1 of grade IV acute GVHD. No patients experienced fatal regimen-related toxicity. Four of the 26 patients developed grade II GVHD or higher; all but 1 responded to therapy with either steroids alone or in addition to ATG. Twenty-two patients had neutrophil recovery at a median of 12 days post transplant (range, 9–31), and 20 patients achieved platelet transfusion independence a median of 13 days post transplant (range, 8–78).

Among the 17 patients with AML or myelodysplastic syndrome (MDS), 10 patients achieved complete remission (<5% bone marrow blast with neutrophil recovery and platelet transfusion independence). Chimerism analysis of the 10 patients achieving complete remission on day 30 revealed that 7 had more than 80% donor cells by either cytogenetics or molecular techniques, 1 patient had autologous reconstitution, and 2 patients were not evaluable for chimerism either because of technical difficulties or lack of difference in enzyme restriction pattern between donor and recipient (syngeneic transplant). At 3 months post transplant, 6 patients remained in remission, of whom 4 had greater than 80% donor cells by similar studies, and by 1 year 1 of the 2 patients in remission remained 100% donor by molecular techniques and the other patient was inevaluable (syngeneic).

The median survival for patients achieving complete remission (CR) was 211 days, with three patients remaining alive at 5, 22, and 23 months, respectively, after transplant. Nine patients have relapsed, of which two were successfully reinduced, one with a second minitransplant and the other with a conventional syngeneic transplant using busulfan/cyclophosphamide conditioning. The median survival of the nonresponding patients was 61 days, with none of them responding to subsequent salvage maneuvers.

Among the nine patients with CML, all patients had hematological recovery of both neutrophil and platelets, with both recipients of unrelated donor cells never showing evidence of donor cell engraftment. Bone marrow on day 30 revealed complete cytogenetic remission in five patients and major cytogenetic remission in another three, with one patient having insufficient metaphases. Five of seven evaluable patients had cytogenetic progression during the first 3 months after transplant; one patient remains in complete cytogenetic remission 6 months post transplant and the other relapsed 9 months post transplant. All five of the progressing patients had immunosuppression withdrawn and received further infusion of donor lymphocytes, with one complete cytogenetic remission reported to date. Overall survival is 53% ± 20% at 2 years, with two of the five surviving patients in complete cytogenetic remission [36]. These results suggest that the combination of fludarabine, idarubicin, and ara-c, although effective in achieving remissions in refractory AML and in CML, is not sufficiently immunosuppressive for unrelated donor cells, and a better antileukemic effect is needed for patients with refractory AML.

We hypothesized that melphalan in combination with purine analogs could potentially improve disease control without undue toxicity in patients considered poor candidates for conventional myeloablative therapies. The rationale for this combination rests upon the broad-spectrum activity of melphalan in a variety of hematological malignancies, its tolerability in patients with myeloma (usually older and many with concurrent medical conditions), the ability of single-agent melphalan in facilitating engraftment of allogeneic progenitor cells, and the observation that fludarabine and 2-CDA have been demonstrated to inhibit the mechanisms of DNA repair after alkylator-induced damage.

From February 1996 to april 1998, a total of 63 patients with a variety of hematological malignancies received melphalan in combination with either 2-CDA or fludarabine [37]. Patient and treatment characteristics are summarized in Table 2. All patients received unmanipulated donor bone marrow or stem cells and FK506–methotrexate combinations for GVHD prophylaxis. Of the 63 patients, 56 demonstrated neutrophil recovery at a median of 14 days

TABLE 2. Patient and treatment characteristics for patients receiving melphalan and purine analog combinations

n	63
Median age, years (range)	49 (22–71)
Time to BMT, days (range)	558 (69–6626)
Diagnosis and stage at BMT	
Acute leukemia	
CR1/CR2	1/2
Untreated first relapse	10
Refractory or other	35
CML	
First chronic phase	2
Accelerated	6
Blast crisis/CP2	3/4
Other	12
Prior regimens	2 (0–9)
Comorbid conditions	
Age > 50 years	29
Prior BMT	17
Poor Organ Function	31
PS 2	12
Donor type	
6/6 Sib	31
5/6 Sib	2
6/6 MUD	30
Preparative regimen	
Fludarabine/melphalan	55
2CDA/melphalan	8

(range, 9–35); 40 patients recovered platelet transfusion independence at a median of 22 days (range, 9–118). All engrafting patients except 1 had documentation of greater than 80% donor cell engraftment by day 30, with 1 instance of autologous reconstitution and 1 case of secondary graft failure. In this group of very poor prognosis patients ineligible for conventional transplant, the 100-day transplant-related mortality (TRM) was 50% (31/63) with 4 of 8 patients in the 2CDA/melphalan arm dying of multiorgan failure, leading us to closing this treatment arm. The overall survival for patients in CR-1 or untreated first relapse was 68% at 1 year, versus 9% for patients with more advanced or refractory disease. We concluded that fludarabine/melphalan combinations can allow engraftment of allogeneic progenitor cells, including cells obtained from matched unrelated donors, that this strategy can produce long-term disease control in patients with hematological malignancies early in the course of their disease, with acceptable risk and toxicity in patients ineligible for conventional myeloablative transplant therapies. Treatment-related mortality from GVHD and disease recurrence remain the most common causes of treatment failure in this patient population.

Minitransplants for Lymphoid Malignancies

The use of allogeneic transplantation is limited in patients with lymphoid malignancies such as CLL or lymphomas because they typically affect older patients. We have evaluated the induction of graft-versus-leukemia as primary therapy for patients with lymphoid malignancies who are considered poor candidates for conventional transplant techniques [38]. Nine patients have been treated, of which eight were older than age 50. All patients with advanced CLL ($n = 5$) or transformed lymphoma ($n = 4$), using one of two preparative regimens (fludarabine/cytoxan or fludarabine/ara-c/platinum). Mixed chimerism was observed in six of nine patients, with percentage of donor cells ranging from 50% to 100% 1 month post transplant. No regimen-related deaths were observed, and four patients achieved complete remission, one after donor lymphocyte infusions.

Summary and Conclusions

The efficacy of graft-versus-leukemia induction to treat relapses after allogeneic progenitor cell transplant in a variety of hematological malignancies suggests that it may be possible to use graft-versus-leukemia (GVL) as primary therapy for these malignancies without the need of myeloablative therapy. This type of strategy should be explored initially in patients considered ineligible for conventional myeloablative therapies either because of age or concurrent medical conditions, and not as a substitute for myeloablative therapy in appropriate candidates for this procedure.

We and others have demonstrated that nonablative chemotherapy using fludarabine combinations is sufficiently immunosuppressive to allow engraftment of allogeneic blood progenitor cells. Patients could then receive graded doses of donor lymphocytes without rejection, to mediate GVL. Ideally, this therapy could be titrated to levels of residual malignant cells using sensitive detection techniques. This novel approach to therapy would reduce the toxicity of the transplant procedure, allow it to be administered more safely to debilitated patients, and possibly extend the use of transplantation to older patients who are not presently eligible for BMT procedures. Other possible indications include treatment of nonmalignant disorders and induction of tolerance for solid organ transplantation.

GVHD remains a major obstacle that needs to be overcome; although a potentially lower level of inflammatory cytokines may be present after nonmyeloablative therapies, fatal GVHD still occurs. Methods to diminish GVHD after allogeneic transplant such as selective T-cell depletion, HSV-TK-transduced lymphocytes, and other nonablative combinations are currently being explored [39–41].

References

1. Gale RP, Champlin RE (1984) How does bone marrow transplantation cure leukemia? Lancet 2:28–30
2. Bortin M, Horowitz M, Gale R, et al (1992) Changing trends in allogeneic bone marrow transplantation for leukemia in the 1980s. JAMA 268:607–612
3. Weiden PL, Sullivan KM, Flournoy N, Storb R, Thomas ED, Seattle Marrow Transplant Team (1981) Antileukemic effect of chronic graft-versus-host disease: contribution to improved survival after allogeneic marrow transplantation. N Engl J Med 304:1529–1532
4. Horowitz MM, Gale RP, Sondel PM, et al (1990) Graft-versus-leukemia reactions after bone marrow transplantation. Blood 75:555–562
5. Sullivan KM, Storb R, Buckner CD, et al (1989) Graft-versus-host disease as adoptive immunotherapy in patients with advanced hematologic neoplasms. N Engl J Med 320:828–834
6. Sullivan KM, Weiden PL, Storb R, et al (1989) Influence of acute and chronic graft-versus-host disease on relapse and survival after bone marrow transplantation from HLA-identical siblings as treatment of acute and chronic leukemia. Blood 73:1720–1728
7. Goldman JM, Gale RP, Bortin MM, et al (1988) Bone marrow transplantation for chronic myelogenous leukemia in chronic phase: increased risk of relapse associated with T-cell depletion. Ann Intern Med 108:806–814
8. Gale RP, Horowitz MM, Ash RC, et al (1994) Identical-twin bone marrow transplants for leukemia. Ann Intern Med 120:646–652
9. Kolb HJ, Schattenberg A, Goldman JM, et al (1995) Graft-vs-leukemia effect of donor lymphocyte transfusions in marrow grafted patients. Blood 86:2041–2050

10. Van Rhee F, Lin F, Cullis JO, et al (1994) Relapse of chronic myeloid leukemia after allogeneic bone marrow transplant: the case for giving donor leukocyte transfusions before the onset of hematologic relapse. Blood 83:3377–3383

11. Mackinnon S, Papadopoulos EB, Carabasi MH, et al (1995) Adoptive immunotherapy evaluating escalating doses of donor leukeocytes for relapse of chronic myeloid leukemia after bone marrow transplantation: separation of graft-versus-leukemia responses from graft-versus-host disease. Blood 86:1261–1268

12. Cullis JO, Jiang YZ, Schwarer AP, Hughes TP, Barrett AJ, Goldman JM (1992) Donor leukocyte infusions for chronic myeloid leukemia in relapse after allogeneic bone marrow transplantation. Blood 79:1379–1381

13. Drobyski WR, Keever CA, Roth MS, et al (1993) Salvage immunotherapy using donor leukocyte infusions as treatment for relapsed chronic myelogenous leukemia after allogeneic bone marrow transplantation: Efficacy and toxicity of a defined T-cell dose. Blood 82:2310–2318

14. Tricot G, Vesole DH, Jagannath S, Hilton J, Munshi N, Barlogie B (1996) Graft-versus-myeloma effect: proof of principle. Blood 87:1196–1198

15. Rondón G, Giralt S, Huh Y, et al (1996) Graft-versus-leukemia effect after allogeneic bone marrow transplantation for chronic lymphocytic leukemia. Bone Marrow Transplant 18:669–672

16. Collins R, Shpilberg O, Drobyski W, et al (1997) Donor leukocyte infusions in 140 patients with relapsed malignancy after allogeneic bone marrow transplant. J Clin Oncol 15:433–444

17. Antin JH (1993) Graft-versus-leukemia: no longer an epiphenomenon. Blood 82:2273–2277

18. Van Bekkum D (1984) Conditioning regimens for marrow grafting. Semin Hematol 21:81–90

19. Deeg H, Liesenring W, Storb R, et al (1998) Long-term outcome after marrow transplantation for severe aplastic anemia. Blood 91:3637–3645

20. Stucki A, Leisenring W, Sandmeier B, Sanders J, Anasetti C, Storb R (1998) Decreased rejection and improved survival of first and second marrow transplant for severe aplastic anemia (a 26-year retrospective analysis). Blood 92:2742–2749

21. Flowers ME, Zanis J, Pasquini R, et al (1996) Marrow transplantation for Fanconi anaemia: conditioning with reduced doses of cyclophosphamide without radiation. Br J Haematol 92:699–706

22. Yu C, Storb R, Mathey B, et al (1995) DLA-identical bone marrow grafts after low dose total body irradiation: Effects of high dose corticosteroids and cyclosporine on engraftment. Blood 86:4376–4381

23. Storb R, Yu C, Wagner J, et al (1997) Stable mixed hematopoietic chimerism in DLA-identical littermate dogs given sublethal total body irradiation before and pharmacological immunosuppression after marrow transplantation. Blood 89:3048–3054

24. Zwaan CM, Van Weel-Sipman MH, Fibbe WE, Oudshoorn M, Vossen JM (1998) Unrelated donor bone marrow transplantation in Fanconi anaemia: the Leiden experience. Bone Marrow Transplant 21:447–453

25. Keating M, Kantarjian H, Talpaz M, et al (1989) Fludarabine: a new agent with major activity against chronic lymphocytic leukemia. Blood 74:19–25

26. Keating MJ, O'Brien S, Robertson LE, et al (1994) The expanding role of fludarabine in hematologic malignancies. Leuk Lymphoma 14(suppl. 2):11–16

27. Redman J, Cabanillas F, Velasquez W, et al (1992) Phase II trial of fludarabine phosphate in lymphoma: an effective new agent in low grade lymphoma. J Clin Oncol 10:790–794

28. Gandhi V, Estey E, Keating MJ, Plunkett W (1993) Fludarabine potentiates metabolism of cytarabine in patients with acute myelogenous leukemia during therapy. J Clin Oncol 11:116–124
29. Li L, Glassman A, Keating M, Stros M, Plunkett W, Yang L (1997) Fludarabine triphosphate inhibits nucleotide excision repair of cisplatin-induced DNA adducts in vitro. Cancer Res 57:1487–1494
30. Estey E, Plunkett W, Gandhi V, Rios MB, Kantarjian H, Keating MJ (1993) Fludarabine and arabinosylcytosine therapy of refractory and relapsed acute myelogenous leukemia. Leuk Lymphoma 9:343–350
31. Delannoy A (1996) 2-Chloro-2°-deoxyadenosine: clinical applications in hematology. Blood Rev 10:148–166
32. Betticher DC, Zucca E, von Rohr A, et al (1996) 2-Chlorodeoxyadenosine (2-CDA) therapy in previously untreated patients with follicular stage III–IV non-Hodgkin's lymphoma Ann Oncol 7(8):793–799
33. Goodman E, Fiedor P, Fein S, et al (1995) Fludarabine phosphate and 2-chlorodeoxyadenosine: immunosuppressive DNA synthesis inhibitors with potential application in islet allo-xenotransplantation. Transplant Proc 27:3293–3295
34. Plunkett W, Sanders P (1991) Metabolism and action of purine nucleoside analogs. Pharmacol Ther 49:239–245
35. Giralt S, Estey E, Albitar M, et al (1997) Engraftment of allogeneic hematopoieetic progenitor cells with purine analog-containing chemotherapy: harnessing graft-versus-leukemia without myeloablative therapy. Blood 89:4531–4536
36. Giralt S, Gajewski J, Khouri I, et al (1997) Induction of graft-vs-leukemia as primary treatment of chronic myelogenous leukemia. Blood 90:418a
37. Giralt S, Cohen A, Mehra R, et al (1997) Preliminary results of fludarabine/melphalan or 2CDA/melphalan as preparative regimens for allogeneic progenitor cell transplantation in poor candidates for conventional myeloablative conditioning. Blood 90:417a
38. Khouri I, Keating MJ, Przepiorka D, et al (1996) Engraftment and induction of GVL with fludarabine-based non-ablative preparative regimen in patients with chronic lymphocytic leukemia. Blood 88(suppl. 1):301a
39. Giralt S, Hester J, Huh Y, et al (1995) CD8+ depleted donor lymphocyte infusion as treatment for relapsed chronic myelogenous leukemia after allogeneic bone marrow transplantation: graft vs. leukemia without graft vs. host disease. Blood 86:4337–4343
40. Bordignon C, Bonini C, Verzeletti S, et al (1995) Transfer of the HSV-tk gene into donor peripheral blood lymphocytes for in vivo modulation of donor anti-tumor immunity after allogeneic bone marrow transplantation. Hum Gene Ther 6:813–819
41. Alyea E, Soiffer R, Canning C, et al (1998) Toxicity and efficacy of defined doses of CD4+ donor lymphocytes for treatment of relapse after allogeneic bone marrow transplant. Blood 91:3671–3680

Ex Vivo Expansion of Human Cord Blood Stem Cells and Genetic Manipulation

KIYOSHI ANDO[1,2], HIROSHI KAWADA[1,2], TAKASHI SHIMIZU[1,3],
TAKASHI TSUJI[5], YOSHIHIKO NAKAMURA[1], MINORU KIMURA[4],
HIROKO MIYATAKE[1], YASUHITO SHIMAKURA[1,2], SHUNICHI KATO[1,3],
and TOMOMITSU HOTTA[1,2]

Summary. To introduce extrinsic genes into hematopoietic stem cells (HSC) by using retrovirus vectors, an efficient system for expanding HSC is required. We have established a novel culture system in which the murine stromal cell line HESS-5 dramatically supports the rapid expansion of cryopreserved cord blood primitive progenitor cells (CB-PPC) in synergy with TPO/FL. Within 5 days of serum-free culture in this system, a 50- to 100-fold increase in CD34+/CD38− cells was obtained; colony-forming units in culture (CFU-C) and mixed colonies (CFU-GEMM) were amplified by 10- to 30 fold and 10- to 20 fold, respectively. To further assess the long-term repopulating ability of those expanded cells, we performed a long-term culture-initiating cells (LTC-IC) assay and SCID-repopulating cells (SRC) assay using CD34+ cells cultured in this system. Within 5 days of culture, 5.1-fold amplification of the LTC-IC was obtained. SRC and their multilineage differentiation were detected in NOD/SCID mice 7 weeks after injection of these cultured cells. This system is further applicable to retrovirus mediated gene transfer to CB-PPC. The transduction efficiency of CD34+ cells was more than 40% when they were infected on HESS-5 monolayer cells. The engraftment of transduced cells in NOD/SCID mice 10 weeks after transplantation were confirmed by the presence of the gene introduced. These results indicate that this xenogeneic coculture system, in combination with human cytokines, can rapidly expand CB stem cells and is applicable to the efficient retrovirus-mediated gene transfer to them.

[1] Research Center for Genetic Engineering and Cell Transplantation
[2] The Department of Internal Medicine
[3] The Department of Pediatrics, and
[4] The Department of Molecular Life Science, Tokai University School of Medicine, Bohseida, Isehara, Kanagawa 259-1193, Japan
[5] Pharmaceutical Frontier Research Laboratory, JT Inc., Kanagawa, Japan

Key words. Cord blood, Hematopoietic stem cell, Stromal cell line, Retrovirus, Gene transfer

Introduction

Hematopoietic stem cells are useful targets for gene therapy [1]. Attempts at human gene therapy using hematopoietic stem cells, however, so far has achieved limited success partly because of their low efficiency of gene transfer [2]. Although the application of fibronectin (FN) or human bone marrow stromal cells has attained a high efficiency of retrovirus-mediated gene transfer to human hematopoietic progenitor cells, more primitive cells such as severe combined immunodeficiency (SCID)-repopulating cells (SRC) were still difficult to transduce. To overcome this limitation, an efficient system for ex vivo expansion of primitive progenitor cells (PPC) is required.

Many studies have been attempted to identify culture condition able to support the expansion of PPC [3–5]. Recently it was reported that the combination of thrombopoietin (TPO) and Flk-2/Flt-3 ligand (FL) maintained production of cord blood primitive progenitor cells (CB-PPC) for more than 6 months in stroma-free liquid culture [6]. Both early-acting cytokines were reported to sustain cell viability and promote proliferation preferentially of a minor subpopulation of CD34+ cells, i.e., CD34+/CD38− cells [7–10]; multipotent progenitor cells and long-term culture-initiating cells (LTC-IC) and SRC are included in this subset [11,12].

On the other hand, the coculture system with bone marrow stromal cells was reported to preserve human PPC quality during ex vivo manipulation [13]. We have established several murine stromal cell lines from bone marrow and spleen [14]. Among these, we discovered a novel hematopoietic-supportive cell line, HESS-5, which effectively supports not only formation of murine granulocyte and macrophage colonies but also proliferation of human CD34+/CD38− cells in the presence of human cytokines [15,16]. Progenitors expanded by this xenogeneic coculture system generate a number of high proliferative potential colony-forming cells (CFC) and differentiate to CD10+/CD19+ B-lymphoid cells.

We demonstrate here the marked supportive effects of HESS-5 cells on proliferation of PPC isolated from cryopreserved CB in synergy with TPO/FL. The number of CD34+/CD38− cells was dramatically increased in serum-free condition for 5 days. LTC-IC and SRC assay of these cells indicated extensive ability to sustain long-term hematopoiesis. This system was further applicable to the efficient retrovirus-mediated gene transfer into CB-PPC. Here we compared the transduction efficiency into CD34+ cells, CFC, and SRC with that of FN.

Materials and Methods

CD34+ Cell Purification and Cryopreservation

CB, collected according to institutional guidelines, was obtained from normal full-term deliveries. CD34+ cell purification, using the MACS immunomagnetic separation system (Miltenyi Biotec, Glodbach, Germany), was done according to the manufacturer's instructions. Ninety-five percent or more of the enriched cells were CD34+ by flow cytometric analysis. Several aliquots were made from the same donor with each containing 10^5 cells and frozen in a medium supplemented with dimethylsulfoxide and FCS (Cell-Banker; Nihon Zenyaku Kohgyo, Fukushima, Japan) using a step-down freezing procedure and placed in liquid nitrogen. Aliquots of frozen samples were thawed before use.

Culture Systems

HESS-5 cells were cultured on the reverse (back) side of the membrane of the culture insert. After obtaining a confluent feeder layer, CD34+ cells were seeded on the upper side of the membrane of the insert where the cytoplasmic villi of HESS-5 cells passed through the etched 0.45-mm pores. Therefore, while HESS-5 cells directly adhered to human hematopoietic cells during culture, expanded cells could easily be harvested without contamination with HESS-5 cells. Serum-free liquid culture was carried out using StemProTM-34SFM (GibcoBRL, Grand Island, NY) supplemented with StemProTM-34 Nutrient Supplement (GibcoBRL) with or without designated cytokines. The final concentrations of cytokines were as follows: TPO, 50 ng/ml; FL-2, 50 ng/ml; IL-3, 20 ng/ml; SCF, 50 ng/ml; GM-CSF, 10 ng/ml; and EPO, 3 U/ml.

LTC-IC Assay

The LTC-IC assay was performed as described by Sutherland et al. [17] with slight modifications. Briefly, bone marrow stromal cells derived from hematologically normal donors were seeded at 10^5 cells per well in 96-well flat-bottomed plates. CD34+ cells purified from CB or those isolated from cultured cells were seeded at limiting dilution on the feeder layer. After 5 weeks of culture, cells were assayed for CFU-C in methylcellulose medium.

SCID Repopulating Cell Assay

The SRC assay was performed as described by Hogan et al. [18] Eight-week-old NOD/shi-scid/scid (NOD/SCID) mice were obtained from the Central Institute for Experimental Animals (Kawasaki, Japan) and maintained in the germfree animal facility located at Tokai University School of Medicine. Preexpanded or postexpanded cells from the cryopreserved aliquots from the same donor were transplanted by tail-vein injection into sublethally irradiated mice (350 cGy). Mice were then killed 6–10 weeks after transplantation, and the bone marrow (BM) from the femurs of each mouse was flushed into RPMI1640 containing 10% fetal calf serum (FCS).

Retrovirus-Mediated Transduction of CD34+ CB Cells

The therapeutic retrovirus vector MFG-GC-GFP and producing cells No. 50 have been previously described [19]. Normal human cord blood CD34+ cells were infected multiple times with virus supernatants supplemented with SCF, IL-3, and IL-6 for 4 days either on HESS-5 cells (HESS-5 method) or fibronectin-coated plates (FN method).

Statistical Analysis

Results are given as mean ± SEM for the three different experiments. Data were compared using analysis of variance. Where significant differences were inferred, sample means were compared using the t-test.

Results

Ex Vivo Expansion of CB-PPC in a Short-Term Serum-Free Culture

Under cytokine-free conditions, HESS-5 cells did not support proliferation of CD34+ cells. However, in the presence of TPO/FL, HESS-5 cells could effectively support proliferation of progenitor cells, especially in the CD34+/CD38− subpopulation; the mean number of CD34+/CD38− cells was approximately 150 times the input number after 14 days of culture. The output of CFU-C and CFU-Mix also was enhanced to 74 and 35 fold, respectively. The synergistic effects of HESS-5 cells and TPO/FL were also studied in serum-free culture for a short duration. After 5 days of culture, the number of progenitor cells, especially CD34+/CD38− cells, was remarkably increased in the presence of TPO/FL (Table 1). An addition of SCF or IL-3 enhanced those effects further. As a result, the mean number of CD34+/CD38− cells was approximately 50- to 100 fold the initial input number by this system. The outputs of CFU-C and CFU-GEMM were increased 10- to 30 fold and 10- to 20 fold, respectively. Three-color flow cytometry revealed that most of the CD34+/CD38− population amplified by this system was negative for most lineage-committed markers but was positive for HLA-DR.

TABLE 1. Evaluation of synergistic effects between HESS-5 cells and TPO/FL-containing cytokines on ex vivo expansion of hematopoietic progenotors in very short-term serum-free culture

Cell population or colony	Cytokine-free	TPO + FL	TPO + FL + SCF	TPO + FL + IL-3
Total number of cells	1.1 ± 0.0	5.3 ± 0.6	14.3 ± 1.2*	17.2 ± 3.9**
CD34(+) cells	1.1 ± 0.0	2.3 ± 0.4	4.8 ± 0.1*	4.9 ± 1.3*
CD34(+)CD38(+) cells	1.1 ± 0.0	1.4 ± 0.2	3.0 ± 0.4*	2.9 ± 0.8
CD34(+)CD38(−) cells	1.1 ± 0.4	48.4 ± 7.6	103.1 ± 10.3*	113.9 ± 21.7**
CFU-C	1.5 ± 0.1	12.4 ± 2.3	34.2 ± 6.7**	33.2 ± 5.3*
CFU-GEMM	0.5 ± 0.1	9.4 ± 2.5	18.3 ± 5.3	11.5 ± 1.2

TPO, thrombopoietin; FL, Flk-2/Flt-3 ligand; SCF, stem cell factor; CFU-C, colong-forming units m culture; CFU-GEMM, mixed colonies.
In the presence or absence of these cytokines, cord blood (CB) CD34(+) cells were cultured with HESS-5 cells under serum-free conditions. HESS-5 cells directly adhered to the CB cells through 0.45-mm pores of the high pore density membrane in the cell culture insert system. Each number at cell populations indicates the mean fold increase + SEM of three different experiments on day 5 of culture. Suitable aliquots of cultured cells were assayed for CFU-C and CFU-GEMM. After 2 weeks of the clonal cell culture, colony scoring was performed and the results represent the fold increase (mean ± SEM of three different experiments).
*$P < 0.05$; **$P < 0.01$, as compared with TPO + FL.
From [16].

LTC-IC and SRC Assay Using the CD34+ Population Isolated from Cells Cultured in Serum-Free Condition

To determine whether the expanded cells could preserve the ability to sustain long-term hematopoiesis, the quantitation of LTC-IC frequency in vitro and the presence of SRC in vivo was assessed. For both assays, CD34+ CB cells cultured in the conditions described and those initially prepared from CB (control samples) were used. When compared with control, the LTC-IC frequency was amplified 5 fold in the CD34+ fraction and 25 fold in CD34+/CD38− fraction (Table 2). More than 50% of human cells were detected in NOD/SCID mice transplanted with the postexpanded CB cells whereas 2% to 5% of cells were detected in mice transplanted with preexpanded cells from the same donor (Fig. 1A; c–f). These cells were differentiated into B lymphocytes, granulocytes, monocytes, erythrocytes, and megakaryocytes (Fig. 1B).

Retroviral Transduction of CD34+ and SRC Using the HESS-5 Coculture System

For gene therapy for Gaucher disease, we have constructed a bicistronic retrovirus vector (pMFG-GC-GFP), which contains both the human glucocere-

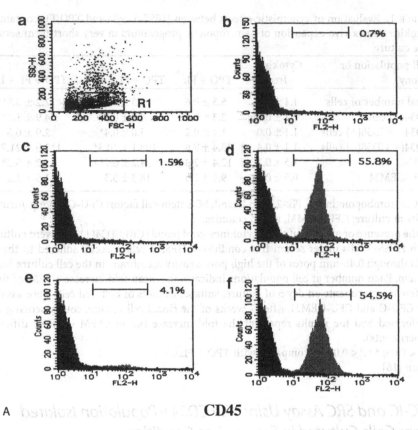

A   **CD45**

FIG. 1A,B. Engraftment and multilineage differentiation of ex vivo expanded CB CD34(+) cells in NOD/SCID mice [16]. Bone marrow from a highly engrafted mouse transplanted with human cord blood cells was stained with various human-specific monoclonal antibodies and analyzed by flow cytometry. A Histogram of human CD45 expression. a Cells with medium to high forward and side scatter (region *R1*) were gated and further analyzed. b Negative control mouse transferred with irradiated human carrier cells. c Mouse transferred with preexpanded 1×10^5 CD34(+) cells from sample 1; 1.5% of the cells present in the murine bone marrow are human. d Mouse transferred with postexpanded cells from 1×10^5 CD34+ cells from sample 1; 55.8% of the cells present in the murine bone marrow are human. e Mouse transferred with preexpanded 1×10^5 CD34+ cells from sample 2; 4.1% of the cells present in the murine bone marrow are human. f Mouse transferred with postexpanded from 1×10^5 CD34+ cells from sample 2; 44.5% of the cells present in the murine bone marrow are human. B Analysis of lineage markers was done on cells within CD45+. Populations of myeloid progenitors (*CD33+, 21.9%*), monocytes (*CD14+, 5.6%*), and megakaryocytes (*CD41+, 7.7%*), erythrocytes (*Gly A+, 4.1%*), B lymphocytes (*CD19+, 6.4%*), and progenitor cells (*CD34+, 5.4%*) were observed

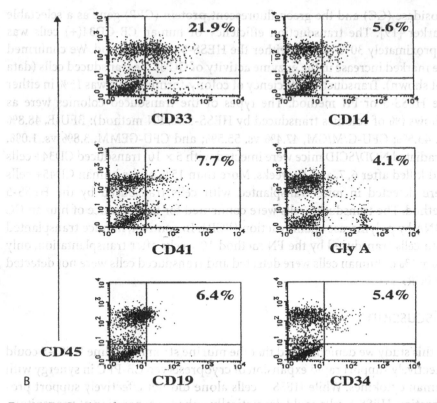

Fig. 1A,B. *Continued*

Table 2. Frequency and fold increase of LTC-IC by the short-term serum-free culture system

Cells	Frequency of LTC-IC		Total fold increase
	Preexpansion	Postexpansion	
CD34(+)	1/102	1/138	5.1 ± 2.3
CD34(+)38(+)	1/215	1/2833	0.2 ± 0.1
CD34(+)38(−)	1/88	1/134	25.3 ± 16.5

LTC-IC, long-term culture-initiating cells.
Same number of cord blood (CB) cells before or after culture in the VST-SFC system were seeded on the human bone marrow stromal cells in 96-well plates. After 5 weeks of culture, cells were assayed for CFU-C, the frequency of wells in which there were no clonogenic cells were determined per the numbers of initial input populations, and the frequency of LTC-IC was calculated. Total fold increase was shown by mean ± SEM of three different experiments.
From [16].

brosidase (GC) and the green fluorescent protein (GFP) gene as a selectable marker [19]. The transduction efficiency of human CB CD34(+) cells was approximately 30%–40% by either the HESS-5 or FN method. We confirmed the marked increase in the enzyme activity of GC of the transduced cells (data not shown). Transduction efficiency of colony-forming cells was 15% in either the HESS-5 or FN method. The types of the transduced colonies were as follows (% of colonies transduced by HESS-5 vs. FN method): BFU-E, 48.8% vs. 43.5%; CFU-G/M/GM, 47.4% vs. 55.5%; and CFU-GEMM, 3.8% vs. 1.0%. Irradiated NOD/SCID mice were injected with 5×10^5 transduced CD34+ cells and killed after 6, 7, and 10 weeks. More than 15% of the human CD45+ cells were detected in mice transplanted with cells transduced by the HESS-5 method. The transduced cells were confirmed for the presence of human GC cDNA by polymerase chain reaction (PCR). In contrast, in mice transplanted with cells transduced by the FN method 10 weeks after transplantation, only 2% ± 1% of human cells were detected and transduced cells were not detected by PCR.

Discussion

In this study we demonstrated that the murine stromal cell line HESS-5 could effectively support rapid expansion of cryopreserved CB-PPC in synergy with human cytokines. While HESS-5 cells alone did not effectively support proliferation, HESS-5 cells could dramatically enhance generation of progenitors, especially CD34+/CD38− cells, in synergy with TPO/FL-2 over a short period. These expanded cells were confirmed to sustain long-term repopulation in LTC-IC and SRC assays. This system was further applicable to retrovirus mediated gene transfer into CB stem cells.

In ex vivo expansion of human PPC, there are several benefits to using a xenogeneic stromal cell line as the feeder: (1) HESS-5 cells can be maintained easily, (2) consistent hematopoietic-supportive effects are repeatedly obtained, and (3) cell differentiation caused by exposure to various factors secreted by stromal cells can be avoided. We previously studied the supportive effects of normal human bone marrow fibroblasts on proliferation of CB-PPC, but variable results were noted because of their induction of differentiation (data not shown). Because HESS-5 cells usually require horse serum for growth and activity [14], we attempted ex vivo manipulation for a very short duration; adequate expansion of CB-PPC was obtained. Most of the expanded CD34+/CD38− cell population expressed HLA-DR. In contrast to previous findings in bone marrow, CB LTC-IC were shown to be present among the CD34+/HLA-DR+ cell fraction [20]. Therefore, the expanded hematopoietic progenitors were expected to sustain long-term hematopoiesis. As a result, the LTC-IC number was increased up to 10 fold within 5 days, and SRC was confirmed in the expanded fraction of cells.

Although hematopoietic stem cells are ideal targets for gene therapy, the transduction efficiency has been very low [2]. As shown here, however, the application of a coculture system with HESS-5 enabled a high level of gene transfer into CD34+ CB cells with SRC ability. These results indicate that the xenogeneic coculture system we have described here, in combination with human cytokines, can rapidly expand CB stem cells and is applicable to the efficient retrovirus-mediated gene transfer to them.

The mechanism of hematopoietic-supportive effects of HESS-5 cells, especially in the CD34+/CD38– cell fraction, remains unknown. Recently, cell-surface molecules identified as delta-like/preadipocyte factor-1 or jagged-1 on stromal cells were reported to show activity of preserving the primitive phenotype of hematopoietic progenitors [21,22]. These novel molecular pathways may play an important role in stem cell regulation in the system we presented here.

Acknowledgment. The authors thank the members of Tokai CBSC Study Group for providing human cord blood. This study was supported by The Japan Society for the Promotion of Science (JSPS) grant no. JSPS-RFTF97I00201 and a Research Grant of The Science Frontier Program from the Ministry of Education, Science, Sports and Culture of Japan.

References

1. Moritz T, Williams DA (1999) Methods for gene transfer: genetic manipulation of hematopoietic stem cells. In: Thomas ED, Blume KG, Froman SJ (eds) Hematopoietic cell transplantation. Blackwell, Malden, p 79
2. Larochelle A, Vormoor J, Hanenberg H, Wang JCY, Bhatia M, Lapidot T, Moritz T, Murdoch B, Xiao XL, Kato I, Williams DA, Dick JE (1996) Identification of primitive human hematopoietic cells capable of repopulating NOD/SCID mouse bone marrow: implications for gene therapy. Nat Med 2:1329
3. Traycoff CM, Kosak ST, Grigsby S, Srour EF (1995) Evaluation of ex vivo expansion potential of cord blood and bone marrow hematopoietic progenitor cells using cell tracking and limiting dilution analysis. Blood 85:2059
4. DiGiusto DL, Lee R, Moon J, Moss K, O'Toole T, Voytovich A, Webster D, Mule JJ (1996) Hematopoietic potential of cryopreserved and ex vivo manipulated umbilical cord blood progenitor cells evaluated in vitro and in vivo. Blood 87:1261
5. Petzer AL, Hogge DE, Landsdorp PM, Reid DS, Eaves CJ (1996) Self-renewal of primitive human hematopoietic cells (long-term-culture-initiating cells) in vitro and their expansion in defined medium. Proc Natl Acad Sci USA 93:1470
6. Piacibello W, Sanavio F, Garetto L, Severino A, Bergandi D, Ferrario J, Fagioli F, Berger M, Aglietta M (1997) Extensive amplification and self-renewal of human primitive hematopoietic stem cells from cord blood. Blood 89:2644
7. Petzer AL, Zandstra PW, Piret JM, Eaves CJ (1996) Differential cytokine effects on primitive (CD34+CD38–) human hematopoietic cells: novel responses to Flt3-ligand and thrombopoietin. J Exp Med 183:2551

8. Kobayashi M, Laver JH, Kato T, Miyazaki H, Ogawa M (1996) Thrombopoietin supports proliferation of human primitive hematopoietic cells in synergy with steel factor and/or interleukin-3. Blood 88:429

9. Haylock DN, Horsfall MJ, Dowse TL, Ramshaw HS, Niutta S, Protopsaltis S, Peng L, Burrell C, Rappold I, Buhring H-J, Simmons PJ (1997) Increased recruitment of hematopoietic progenitor cells underlies the ex vivo expansion potential of FLT3 ligand. Blood 90:2260

10. Borge OJ, Ramsfjell V, Cui L, Jacobsen SEW (1997) Ability of early acting cytokines to directly promote survival and suppress apoptosis of human primitive CD34+CD38– bone marrow cells with multilineage potential at the single-cell level: key role of thrombopoietin. Blood 90:2282

11. Hao QL, Shah AJ, Thiemann FT, Smogorzewska EM, Crooks GM (1995) A functional comparison of CD34+CD38– cells in cord blood and bone marrow. Blood 86:3745

12. Bhatia M, Bonnet D, Kapp U, Wang JCY, Murdoch B, Dick JE (1997) Quantitative analysis reveals expansion of human hematopoietic repopulating cells after short-term ex vivo culture. J Exp Med 186:619

13. Breems DA, Blokland EAW, Siebel KE, Mayen AEM, Engels LJA, Ploemacher RE (1998) Stroma-contact prevents loss of hematopoietic stem cell quality during ex vivo expansion of CD34+ mobilized peripheral blood stem cells. Blood 91:111

14. Tsuji T, Ogasawara H, Aoki Y, Tsurumaki Y, Kodama H (1996) Characterization of murine stromal cell clones established from bone marrow and spleen. Leukemia (Basingstoke) 10:803

15. Tsuji T, Nishimura Y, Watanabe Y, Hirano D, Nakanishi S, Mori KJ, Yatsunami K (1999) A murine stromal cell line promotes the expansion of CD34high+ primitive progenitor cells isolated from human umbilical cord blood in combination with human cytokines. Growth Factors 16:225

16. Kawada H, Ando K, Tsuji T, Shimakura Y, Nakamura Y, Chargui J, Hagihara M, Itagaki H, Shimizu T, Inokuchi S, Kato S, Hotta T (1999) Rapid ex vivo expansion of human umbilical cord hematopoietic progenitors using a novel culture system. Exp Hematol 27:904

17. Sutherland HJ, Lansdorp PM, Henkelman DH, Eaves AC, Eaves CJ (1990) Functional characterization of individual human hematopoietic stem cells cultured at limiting dilution on supportive marrow stromal layers. Proc Natl Acad Sci USA 87:3584

18. Hogan CJ, Shpall EJ, McNulty O, Dick JE, Shultz LD, Keller G (1997) Engraftment and development of human CD34(+)-enriched cells from umbilical cord blood in NOD/LtSz-scid/scid mice. Blood 90:85

19. Shimizu T, Ando K, Kimura M, Miyatake H, Inokuchi S, Takakura I, Migita M, Shimada T, Kato S (1998) A simple and efficient purification of transduced cells by using green fluorescent protein gene as a selection marker. Acta Paediatr Jpn 40:575

20. Traycoff CM, Abboud MR, Laver J, Brandt JE, Hoffman R, Law P, Ishizawa L, Srour EF (1994) Evaluation of the in vitro behavior of phenotypically defined populations of umbilical cord blood hematopoietic progenitor cells. Exp Hematol 22:215

21. Moore KA, Pytowski B, Witte L, Hicklin D, Lemischka IR (1997) Hematopoietic activity of a stromal cell transmembrane protein containing epidermal growth factor-like repeat motifs. Proc Natl Acad Sci USA 94:4011

22. Varnum-Finney B, Purton LE, Yu M, Brashem-Stein C, Flwers D, Staats S, Moore KA, Le Roux I, Mann R, Gray G, Artavanis-Tsakonas S, Bernstein ID (1998) The Notch ligand, jagged-1, influences the development of primitive hematopoietic precursor cells. Blood 91:4084

Part 3
International Collaboration in Hematopoietic Stem Cell Transplantation

Part 3
International Collaboration
in Hematopoietic Stem
Cell Transplantation

Unrelated Donor Stem Cell Transplantation: A Registry Perspective

CRAIG W.S. HOWE, CRAIG KOLLMAN, and JANET HEGLAND

Worldwide, more than 5 million volunteers have been HLA-typed as potential hematopoietic stem cell (HSC) donors for patients who otherwise would not have the option of transplant therapy. The organization and responsibilities of the more than 43 registries vary significantly, generally dependent upon their financial support.

Until recently HSC donor recruitment has embraced a "come one, call all" philosophy. However, (1) the overall lower likelihood of a match for patients for minority populations, (2) the increasing redundancy of donors with common HLA phenotypes, (3) the better understanding of HLA population genetics, and (4) the need to develop cost-effective recruitment all argue strongly for developing scientifically based, internationally coordinated donor recruitment.

Concurrent with the rapid growth of registries, a revolution in HLA typing technology is occurring. As serology is supplanted by DNA-based methods and as clinical research identifies the importance of matching at "new" HLA loci (e.g., HLA-C and HLA-DQ), registries must develop consensus methods for storing and reporting the voluminous HLA data and for modifying the search algorithm, thereby redefining "matched unrelated donor."

The retrospective analysis of survival data from transplants facilitated by one registry, the National Marrow Donor Program, has found significantly better survival among patients receiving marrow from younger donors. Survival of patients transplanted for severe aplastic anemia or myelodysplastic syndrome appears to be more dependent upon having a serologically matched unrelated donor than is the case for patients transplanted to treat acute or chronic leukemia. The collection, analysis, and publication of transplant out-

National Marrow Donor Program, National Coodinating Center, 3433 Broadway Street N.E., Suite 500, Minneapolis, MN 55413, USA

comes by donor registries serve both to advance the field and as an obligation to volunteer donors.

Because donor registries serve patients while attempting to minimize donor risk, they must confront ethical questions. As increased requests are made of donors, it can be anticipated that the principles of justice, autonomy, parenteralism, and medical futility will increasingly be the subject of registry deliberations.

It is apparent that registries of unrelated HSC donors are not simply lists of HLA-typed individuals, but rather participants in shaping the future of transplantation.

European Bone Marrow Donor Registries: Collaboration and Individualism

Torstein Egeland

Summary. The first registry of HLA-typed volunteer bone marrow donors made available for patients residing abroad was The Anthony Nolan Bone Marrow Trust founded in 1974 in London, England. Several years passed before other countries established registries. Later, because of the growing number of donors and need to facilitate donor search processes and tracking, the European Bone Marrow Transplantation Group (EBMT) took the initiative to established Bone Marrow Donors Worldwide (BMDW), and BMDW was generated in 1988 in Leiden, The Netherlands, under the auspices of the Europdonor Foundation. As of November 25, 1998, there are 5.4 million donors in BMDW from 59 registries and cord blood banks in 34 countries. 2.1 million donors, of whom 39% are AB- and DR typed, are from 43 registries and cord blood banks in 25 European countries. In Europe, search processes, donor activation, and confirmatory typing requests take place either through telematic systems (EDS, developed by France Greffe de Moelle in Paris, France, or EMDIS, initially developed by registries in England, France, and Germany) or by direct communication. The large number of individual registries and multiple transplant centers across Europe underlines the need for a common telematic system. Today, several registries have to be addressed in a unique way to order certain services, some services may not be available, and reporting modes vary. Individual policies, varying resources, and national regulations make it a great challenge to integrate registries and transplant centers into a common system. On the other hand, the current "open" system allows a high degree of flexibility and individual service, which for instance may ease final donor workup and (re)transplantation planning in complicated or urgent settings.

The Norwegian Bone Marrow Donor Registry, Institute of Transplantation Immunology, Rikshospitalet, The National Hospital, N-0027 Oslo, Norway

Key words. Bone marrow, Bone marrow donor registry, Bone marrow transplantation, Unrelated donor, HLA

Introduction

Allogeneic bone marrow transplantation was established in the 1960s as a mean to treat various forms of leukemia. Later, allogeneic bone marrow transplantation was also used for treatment of other severe hematological diseases, and certain inborn metabolic and immunodeficient disorders were also included for this treatment. The number of patients who could be offered allogeneic transplantation was limited by the availability of human leukocyte antigen- (HLA-) identical family donors, usually siblings, and only about 30%–35% of the patients could be transplanted. Thus, there was a great need for registries with volunteer donors, and when improved HLA typing techniques, including high-resolution genomic HLA class II typing, became available, transplantation with unrelated donors became an option.

The Anthony Nolan Bone Marrow Trust, established in 1974 in London, England, was the first bone marrow donor registry established to offer bone marrow to patients residing abroad [1]. In 1981, The Anthony Nolan Bone Marrow Trust made history by providing bone marrow transplantation from an unrelated donor to a patient in another country [1]. This event was the start of an exchange of bone marrow harvests across international boundaries. The Anthony Nolan Bone Marrow Trust was followed in 1986 by the establishment of the National Marrow Donor Program (NMDP) in Minneapolis, MN, USA, and in 1987 by the French bone marrow donor registry, France Greffe de Moelle in Paris, the British Bone Marrow Donor Registry in Bristol, England, and the Israeli Donor Bank Registry in Jerusalem, Israel. The establishment of these first registries, the development of reliable techniques for genomic high-resolution HLA class II typing, and several reports of successful transplantation with HLA-matched unrelated donors [2–6] laid the ground for the establishment in the late 1980s and early 1990s of bone marrow donor registries throughout Europe (Table 1) [7].

Since the establishment of bone marrow donor registries in the late 1980s and early 1990s, the number of donors in Europe has increased constantly (Fig. 1). This increase parallels the worldwide increase in number of donors [8]. In Europe, the number of German donors has shown the fastest increase. The number of European donors, including the number of ABDR- and AB-typed donors, as of November 25, 1998, is listed in Table 2.

TABLE 1. European registries of HLA-typed volunteer bone marrow donors according to year of establishment

Location of registries:		
Country	City	Year established
UK (England)	London	1974
France	Paris	1987
UK (England)	Bristol	1987
Israel	Jerusalem	1987
Belgium	Brüssels	1988
Greece	Athens	1988
Netherlands	Leiden	1988
Switzerland	Bern	1988
Austria	Vienna	1989
UK (Wales)	Pontyclun	1989
Ireland	Dublin	1989
Italy	Genoa	1989
Greece	Thessaloniki	1990
Hungary	Budapest	1990
Norway	Oslo	1990
Czech Rep.	Prague	1991
Denmark	Århus	1991
Germany	Ulm	1991
Spain	Barcelona	1991
Czech Rep.	Plzen	1992
Finland	Helsinki	1992
Slovenia	Ljubliana	1992
Sweden	Huddinge	1992
Poland	Wroclaw	1993
Cyprus	Nicosia	1994
Croatia	Zagreb	1996
Portugal	Lisbon	1996
Israel	Petach Tikva	1997
Russia	Moscow	
Slovakia	Bratislava	

Data obtained from [7], as well as from individual registries.

European Interregistry Collaboration

There are 31 bone marrow donor registries in 25 countries in Europe. A few European countries have more than one registry (Table 2). In principle, no transnational body has shaped the formation and organization of individual European registries. Further, individual registries have to function in compliance with national laws and regulations. Thus, the registries have come to their own conclusions and policies regarding establishment and funding,

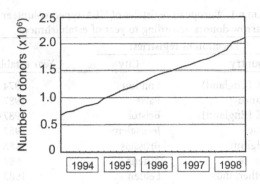

Fig. 1. The growth of number of donors in European registries since the end of 1993 to the end of 1998. (Data were kindly compiled by Jack N.A. Bakker, Bone Marrow Donors Worldwide)

donor recruitment strategies, HLA typing facilitation, donor information and filing, confidentiality, minimum infectious disease marker analyses, donor follow-up, etc.

However, European interregistry collaboration has also been generated to improve donor search and tracking strategies as well as means of reporting donor searches and typing results. These efforts have in particular been financially supported by the European Commission, but some registries have also gained govermental and/or private funding to improve European and international collaboration.

Europdonor Secretariat

In 1989, France Greffe de Moelle obtained a grant from the European Commission to study the flow of communication between European registries and the capacity of developing a telematic system [9]. As a consequence, France Greffe de Moelle set up a donor search and reporting telematic system named Europdonor Secretariat (EDS) with a centralized computer in Paris and communication via telephone lines and modems to computers provided to participating European registries. Today, 17 European and 2 non-European registries participate in the EDS system (Table 3). Several thousand donor search requests, donor DR typing reports, and cancellations of previously activated donors take place annually; e.g., The Norwegian Bone Marrow Donor Registry received 2617 search requests through EDS in 1997. EDS communication occurs on workdays, and the response time is on average 4 days [9].

TABLE 2. Number of HLA-typed volunteer bone marrow donors in European registries as of November 25, 1998, sorted by country: number of donors typed for AB and DR and for AB only

Location of registries:		Total number of donors	Typed for AB and DR	Typed for AB only	Percentage DR typed
Country	City				
Austria	Vienna	33 167	12 800	20 367	39
Belgium	Brussels	50 370	33 128	17 242	66
Croatia	Zagreb	18	18	0	100
Cyprus	Nicosia	790	413	377	52
Czech Rep.	Plzen	12 924	4 076	8 848	32
Czech Rep.	Prague	8 232	950	7 282	12
Denmark	Århus	7 683	4 562	3 121	59
Denmark	Copenhagen	898	46	852	5
Finland	Helsinki	16 175	15 703	472	97
France	Paris	96 797	55 779	41 018	58
Germany	Ulm	1 075 457	277 126	798 331	26
Greece	Athens	917	696	221	76
Greece	Thessaloniki	858	304	554	35
Hungary	Budapest	1 294	342	952	26
Ireland	Dublin	11 314	10 985	329	97
Israel	Jerusalem	12 885	1 857	11 028	14
Israel	Petach Tikva	11 738	121	11 617	1
Italy	Genoa	230 861	68 949	161 912	30
Netherlands	Leiden	25 532	24 863	669	97
Norway	Oslo	16 649	14 288	2 361	86
Poland	Wroclaw	69	23	46	33
Portugal	Lisbon	497	49	448	10
Russia	Moscow	8 225	808	7 417	10
Slovakia	Slovakia	189	4	185	2
Slovenia	Ljubljana	292	193	99	66
Spain	Barcelona	27 008	10 304	16 704	38
Sweden	Huddinge	40 258	13 311	26 947	33
Switzerland	Bern	14 405	13 480	925	94
UK (England)	London	287 954	167 600	120 354	58
UK (England)	Bristol	89 739	60 331	29 408	67
UK (Wales)	Pontyclun	19 547	19 547	0	100
Total		2 102 742	812 656	1 290 086	39

Data obtained from [8].

European Marrow Donor Information System

Because of the rapidly growing number of new donors as well as the continuous introduction of new HLA alleles, it became necessary to establish a completely automatic telematic system to speed up and facilitate donor search and tracking in Europe. Based on grants from the European Commission, a system named European Marrow Donor Information System (EMDIS) was generated

TABLE 3. Location of registries participating in the Europdonor Secretariat
(EDS) telematic system

Country	City
Austria	Vienna
Belgium	Brussels
Czech Rep.	Prague
UK (England)	London
France	Paris
Netherlands	Leiden
Hungary	Budapest
Germany	Ulm
Greece	Athens
Ireland	Dublin
Italy	Genoa
Norway	Oslo
Portugal	Lisbon
Spain	Barcelona
Sweden	Huddinge
Switzerland	Bern
UK (Wales)	Pontyclun
Australia[a]	Sydney
USA (C.R.I.R.)[a]	Worcester, MA

[a] Non-European registries participating
in the EDS system.
Data obtained from the EDS software.

in the 1990s by a joint effort of The Anthony Nolan Bone Marrow Trust, France Greffe de Moelle, and The German Registry of Bone Marrow Donors, and in collaboration with several other European registries [10]. Twelve European countries now participate in the EMDIS system (Table 4) [10].

Today, the EDS and the EMDIS systems are both functioning. In some registries, they work in parallel; in others, only EDS is implemented. In addition, a few registries are not participating in either system. Finally, in addition to the mentioned European telematic systems, The German Registry of Bone Marrow Donors and the German donor centers have, in response to the rapidly growing German donor file, generated a national communication system labeled GERMIS designed to increase the speed and efficiency of donor searches in Germany [11].

Bone Marrow Donors Worldwide

In 1988, Bone Marrow Donors Worldwide (BMDW) was established by an initiative of the Immunobiology Working Party of the European Bone Marrow Transplantation Group (EBMT) [8]. The centralized data file was established

TABLE 4. Location of registries participating in the European Marrow Donor Information System (EMDIS) telematic system

Country	City
Austria	Vienna
Belgium	Brussels
Denmark	Århus
England	London
France	Paris
Germany	Ulm
Greece	Athens
Netherlands	Leiden
Ireland	Dublin
Italy	Genoa
Spain	Barcelona
Switzerland	Bern

Data obtained from [10].

in Leiden, The Netherlands, under the auspices of the Europdonor Foundation, with the task of compiling HLA antigens and haplotypes from donors in registries throughout the world. Registries in Austria, Belgium, England, France, Germany, Italy, the Netherlands, and the United States took part in the establishment.

Initially, the BMDW data files were updated every 3 months, and the files were distributed by books. The first book contained the HLA antigens of 155 000 volunteer bone marrow donors [8]. The HLA data were soon made available through floppy disks, which also included search softwares, and the data file updating became more frequent. Today, the HLA types of more than 5.4 million HLA-typed donors and umbilical cord blood cryopreservates from all over the world are filed in the BMDW data bank, and HLA match and mismatch search programs, as well as counseling provided by the transplant immunologists and computer technologists at Europdonor Foundation, are made available through the BMDW home page on Internet [8].

European Registries and Individualism

Given the individual characteristics of European countries, different national laws and regulations, diverse languages, cultures, traditions, political situations and priorities, it has been a great task to establish the collaborations seen today, for instance, EDS and EMDIS. However, a number of tasks remain if all European registries are to be integrated in a common telematic system that takes care of every aspect related to the activities of donor, transplant, and collections centers, such as search, tracking, requesting, reporting, preharvest workup, and follow-up. At present, the World Marrow Donor Association (WMDA), an international organization based in Leiden, the Netherlands, aiming at establishing guidelines and recommendations for registries and transplant centers throughout the world, is one of the leading bodies to make donor search, HLA typing and reporting, and workup processes easy, fast, accurate and standardized. Representatives from European registries along with representatives from registries on all continents take active part in these efforts, and a number of achievements has already been reached.

European transplant centers, however, still must interact with individual European registries very much on a center-to-registry basis, at least after the preliminary search process. This status creates a number of individual routines with which one must become familiar with to ensure rapid and accurate requests and reports. For instance, some European registries require the use of their own forms to order DR typing or blood sample shipment, but other registries do not. Some registries offer genomic HLA class II typing; others do not. Workup routines vary between registries, and usage of different forms may again be required. Policies and practice regarding HLA matching criteria, donor–patient confidentiality, minimum preharvest donor infectious disease marker analyses, charges, and fees also vary among registries.

Donor Recruitment and HLA Typing

Different strategies have been established in European registries with respect to donor recruitment and HLA typing of new donors. Some donor centers recruit donors through mass media attention, drives in big companies, and local communities, etc., and some recruit only blood donors. Among the advantages of mass media attention and large drives is the fact that many donors can be added to the donor file within a short period of time. Fewer donors can be recruited when restricted to blood donors only, but there is probably less attrition. Also, many blood banks take advantage of using the HLA-typed donors as platelet donors.

As can be seen from Table 2, the relative number of AB- and ABDR-typed donors varies between countries. The number of ABDR-typed donors has a significant impact on donor availability and marrow donation. Thus, in parallel with earlier observations [1,12], data from European bone marrow donor registries indicate that there is a highly significant correlation between the

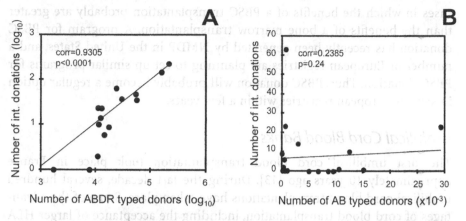

FIG. 2A,B. The number of international marrow donations in 1997 is highly significantly correlated with the number of ABDR-typed donors (A). The number of international donations is, when analyzing registries with fewer than 50 000 donors, not correlated with the number of donors typed for AB only (B). (The data were obtained from the Donor Registries Annual Report 1997, World Marrow Donor Association, Donor Registries Working Group, with permission from individual registries; the report is available through national registries)

number of international donations and ABDR-typed donors (Fig. 2A). For registries with fewer than 50000 donors, there is no correlation with the number of donors typed for AB only (Fig. 2B). These analyses are based on international donations because more donations in Europe take place across than within national borders, and because international donations probably reflect donor availability better than domestic ones as the result of individual national donor evaluation strategies.

Alternative Donors or Stem Cell Compartments in Europe

Peripheral Blood Stem Cells and Unrelated Donors

Although peripheral blood stem cell (PBSC) transplantation is used in an increasing number of allogeneic transplantations from an HLA-identical family donor, donation of PBSCs from unrelated donors has not yet become a routine. Although a significant number of European registries offer PBSCs for a second donation to a patient in the need of a retransplantation, PBSC transplantation with unrelated donors is, except for a very few donor centers, not offered [1]. Exceptions might be well-qualified and well-documented cases in which the benefits of a PBSC transplantation probably are greater than the benefits of a bone marrow transplantation. A program for PBSC donation has recently been generated by NMDP in the United States, and a number of European registries are planning to set up similar programs for PBSC donation. Thus, PBSC donation will probably become a regular option in several European registries within a few years.

Umbilical Cord Blood Banks

The first umbilical cord blood transplantation took place in France approximately 10 years ago [13]. During the last decade, several hundred umbilical cord blood transplantations have taken place. Some of the advantages of cord blood transplantation, including the acceptance of larger HLA disparities than are otherwise accepted in unrelated bone marrow transplantations [14], are that partially matched cord blood cryopreservates could be used to transplant patients with infrequent haplotypes. Umbilical cord blood banks have been or are being set up in several cities around the world, including 12 cities in Europe, 4 in the United States, and 1 in Australia (Table 5). Finally, a European registry named Eurocord collects and provides information about the outcome of cord blood transplantations [15], and an international telematic system for cord blood searches labeled Netcord has been established [16].

Conclusions

Allogeneic bone marrow transplantation with HLA-identical donors are well established, and certain degrees of HLA class I mismatches may also be accepted, particularly in younger patients [17,18]. Successful transplantations with related donors with one fully mismatched HLA haplotype have also been reported, provided that T-cell depletion and transfusion of large doses of CD34+ hematopoietic stem cells are applied [19], and this may also open new possibilities for transplantation with unrelated donors. Several studies are ongoing to further elucidate what kinds of HLA mismatches are accepted when unrelated donors are used. In addition, transplantation with PBSC and with HLA-mismatched umbilical cord blood are new options. Finally, new protocols for "minor" transplantations for the induction of mixed himerisms, ex vivo stem cell expansion, particularly for umbilical cord blood stem cells, and treatment of leukemias or leukemic relapses with donor lymphocytes are all challenges to today's bone marrow donor registries and umbilical cord blood banks. The main goal is to allow all patients who need an allogeneic stem cell transplantation, irrespective of the infrequency of their HLA haplotypes, to be transplanted with a safe protocol.

TABLE 5. Number of cryopreserved umbilical cord blood samples in European and non-European cord blood banks (as of November 25, 1998)

Country	City	Total number of cord samples	Typed for AB and DR	Typed for AB only	Percent DR typed
Europe:					
Austria	Vienna	3	3	0	100
Belgium	Brussels	986	972	14	99
Belgium	Leuvens	516	516	0	100
Czech Rep.	Prague	63	59	4	94
Germany	Düsseldorf	2288	2281	7	100
Germany	Ulm	266	265	1	100
France	Paris	717	675	42	94
Israel	Jerusalem	215	208	7	97
Italy	Milan	3212	3212	0	100
Netherlands	Leiden	328	327	1	100
Spain	Barcelona	2005	2005	0	100
United Kingdom	Bristol	1928	1928	0	100
Total cord blood in Europe		12527	12451	76	99
Outside Europe:					
Australia	Sydney	1114	1114	0	100
USA	Worcester MA	1230	1230	0	100
USA	New York NY	6795	6657	138	98
USA	Paramus NJ	347	347	0	100
USA	St. Louis MO	3036	3035	1	100
Total cord blood outside Europe		12522	12383	139	99
Total cord blood, worldwide		25049	24834	215	99

Data obtained from [8].

Acknowledgments. Thanks are due to European registries for permission to use data compiled in the Donor Registries Annual Report 1997 by World Marrow Donor Association, Donor Registries Working Group. Comments and suggestions from Susan A. Cleaver, Jon J. van Rood, Machteld Oudshoorn, and Collette Raffoux are gratefully acknowledged. Finally, thanks to Jack N.A. Bakker, who compiled data for European registries 1993–1998, and to Lars Mørkrid for statistical analyses.

References

1. Cleaver SA (1997) The World Marrow Donor Association. In: 1996 Annual Report and Accounts. The Anthony Nolan Bone Marrow Trust, London, pp 39–40
2. Beatty PG, Hansen JA, Anasetti C, Sanders J, Buckner CD, Storb R, Thomas ED (1989) Marrow transplantation from unrelated HLA-matched volunteer donors. Transplant Proc 21:2993–2994

3. Hows J, McKinnon S, Brookes P, Kaminski E, Bidwell J, Bradley B, Batchelor J, Goldman J (1989) Matched unrelated donor transplantation. Transplant Proc 21:2923–2925
4. Ash RC, Casper JT, Chitambar CR, Hansen R, Bunin N, Truitt RL, Lawton C, Murray K, Hunter J, Baxter-Lowe LA, Gottschall JL, Oldham K, Anderson T, Camitta B, Menitove J (1990) Successful allogeneic transplantation of T-cell-depleted bone marrow from closely HLA-matched unrelated donors. N Engl J Med 322:485–494
5. McGlave PB, Beatty P, Ash R, Hows JM (1990) Therapy for chronic myelogenous leukemia with unrelated donor bone marrow transplantation: results in 102 cases. Blood 75:1728–1932
6. Kernan NA, Bartsch G, Ash RC, Beatty PG, Champlin R, Filipovich A, Gajewski J, Hansen JA, Henslee-Downey J, McCullough J, McGlave P, Perkins HA, Phillips GL, Sanders J, Stroncek D, Thomas ED, Blume KG (1993) Analysis of 462 transplantations from unrelated donors facilitated by the National Marrow Donor Program. N Engl J Med 328:593–602
7. Egeland T (1997) European bone marrow donor registries: a survey. In: Rossi U, Massaro AL, Sciorelli G (eds) Proceedings of the ISBT 5th Regional (4th European) Congress, Venezia, 2nd–5th July 1995. Edizioni SIMTI, Milan, pp 803–808
8. Bone Marrow Donors Worldwide Internet address: www.bmdw.org
9. Raffoux F (1997) Registries of volunteer donors European Organization. In: Rossi U, Massaro AL, Sciorelli G (eds) Proceedings of the ISBT 5th Regional (4th European) Congress, Venezia, 2nd–5th July 1995. Edizioni SIMTI, Milan, pp 789–792
10. EMDIS Internet address: www.zkrd.uni-ulm.de/emdis/index.htm
11. Annual report (1997) German National Bone Marrow Donor Registry, Ulm, PP 16–20
12. Hurley CK, Wade J, Oudshoorn M, Middleton D, Kukuruga D, Navarrete C, Christiansen F, Hegland J, Ren E-C, Andersen I, Cleaver SA, Brautbar C, Raffoux C (1999) A special report: histocompatibility testing guidelines for hematopoietic stem cell transplantation using volunteer donors. Tissue Antigens 53:394–406
13. Gluckman E, Broxmeyer HE, Auerbach AD, Friedman HS, Douglas GW, Devergie A, Esperou H, Thierry D, Socie G, Lehn P, Cooper S, English D, Kurtzberg J, Bard J, Boyse EA (1989) Hematopoietic reconstitution in a patient with Fanconi's anemia by means of umbilical-cord blood from an HLA-identical sibling. N Engl J Med 321:1174–1178
14. Rubenstein P, Carrier C, Scaradavou A, Kurtzberg J, Adamson J, Migliaccio AR, Berkowitz RL, Cabbad M, Dobrila NL, Taylor PE, Rosenfield RE, Stevens CE (1998) Outcomes among 562 recipients of placental-blood transplants from unrelated donors. N Engl J Med 339:1565–1577
15. Gluckman E, Rocha V, Chastang C (1998) Cord blood banking and transplant in Europe. Eurocord. Bone Marrow Transplant 22(suppl 1):68–74
16. Netcord Internet address: www.kmsz.uni-duesseldorf.de/NETCORD
17. Petersdorf EW, Gooley TA, Anasetti C, Martin PJ, Smith AG, Mickelson EM, Woolfrey AE, Hansen JA (1998) Optimizing outcome after unrelated marrow transplantation by comprehensive matching of HLA class I and II alleles in the donor and recipient. Blood 92:3515–3520
18. Sasazuki T, Juji T, Morishima Y, Kinukawa N, Kashiwabara H, Inoko H, Yoshida T, Kimura A, Akaza T, Kamikawaji N, Kodera Y, Takaku F (for the Japan Bone Marrow Donor Program) (1998) Effect of matching of class I HLA alleles on clinical outcome after transplantation of hematopoietic stem cells from an unrelated donor. N Engl J Med 339:1177–1185

19. Aversa F, Tabilio A, Velardi A, Cunnigham I, Terenzi A, Falzetti F, Ruggeri L, Barbabietola G, Aristei C, Latini P, Reisner Y, Martelli MF (1998) Treatment of high-risk acute leukemia with T-cell-depleted stem cells from related donors with one fully mismatched HLA haplotype. N Engl J Med 339:1186–1193

Asian Bone Marrow Donor Registries: Their Roles in International Cooperation

SHINICHIRO OKAMOTO

Although large registries of volunteer, unrelated marrow donors are now operating around the world, potential recipients are frequently unable to identify suitable donors. These searching patients may benefit from improved international coordination and cooperation, a better understanding of the HLA system and the requirements for HLA matching, and the development of alternative hematopoietic stem cell sources. In the United States, the National Marrow Donor Program has an established role in facilitating and promoting local cooperation, and in Europe, the European Group for Blood and Marrow Transplantation serves that function. In Asia several registries have been established and are operating nationally and internationally. However, the association of registries in Asia has not been established. The purposes of this association include (1) to promote recruitment of volunteer donors of all major and genetically distinct ethnic groups in the Asian region and (2) to provide resources of information and expertise for countries wishing to establish a donor registry for unrelated transplantation. Furthermore, the association of Asian registries is particularly important to establish local, ethical, medical, technical, and financial guidelines for the recruitment and testing of donors, including HLA typing, collection and transport of marrow, and safety monitoring. This association would be a crucial step in the global collaboration of all marrow registries.

Japan Marrow Donor Program, Keio University School of Medicine, 35 Shinanomach, Shinjuku-ku, Tokyo 160-8582, Japan

Cord Blood Stem Cell Transplantation: Banking and Clinical Results

V. Rocha and E. Gluckman

Summary. Since the demonstration that the number of hematopoietic stem cells contained in one cord blood was sufficient for engrafting children and adults, cord blood banking has developed worldwide. Cord blood banking for allogeneic unrelated and related transplants has several advantages including availability of this source of stem cells, low viral infection rate at birth, speed of search, and the possibility to collect cord blood in ethnic groups under-represented in bone marrow donors registries. Another possible advantage includes the low risk of acute graft-versus-host disease (GVHD), even with some degree of HLA mismatch. Currently, more than 25000 units are available for transplantation. To develop and evaluate cord blood transplant (CBT) results, the European Blood and Marrow Transplantation group (EBMT) has organized a concerted action, the Eurocord group. The objectives of Eurocord are to study the properties of hematopoietic progenitors and gene transfer in cord blood; to study the immune function of cord blood lymphocytes; to coordinate and facilitate the exchange of sera and cells from donor and recipients of cord blood transplants; to establish a European registry of patients treated by cord blood transplants; and to design protocols comparing cord blood transplants with alternative conventional blood and bone marrow hematopoietic stem cell transplants. Recently, Netcord, a nonprofit organization, was created for establishing criteria of qualification and accreditation of cord blood banks, and a network was set up between cord blood banks for facilitating donor search. The purpose is to ensure the quality of the cord blood units obtained for transplantation and to aid transplant centers in finding suitable units to increase the number of cord blood transplants (CBT) throughout the world. Briefly, recent analysis of the clinical results in 255 CBTs performed in

Bone Marrow Transplant Unit and Clinical Research Unit, Hopital Saint Louis, 1 Ave. Claude Vellefaux, 75475 Paris, Cedex 10, France

90 countries has shown that factors associated with better survival in related and unrelated transplants were younger age, diagnosis with better results in children with inborn errors, and acute leukemia in first and second remission. High numbers of nucleated cells in the transplant and recipient-negative CMV serology were also favorable risk factors for survival. The most important factor influencing engraftment was the number of cells infused.

Key words. Cord blood transplantation, Unrelated, Related, Banking

Introduction

Allogeneic hematopoietic stem cell transplantation (HSCT) has been used to treat thousands of patients, both adults and children, with life-threatening hematological diseases. The main limitations of allogeneic bone marrow transplantation are the lack of suitable HLA-matched donors and the complications of graft-versus-host disease (GVHD) associated with HLA disparities. In the absence of a suitable HLA-identical sibling donor, alternative donors such as mismatched related or matched unrelated donors are searched. Despite bone marrow donor registries including more than 5 million bone marrow donors worldwide, some patients cannot be transplanted because of the lack of an HLA-identical donor.

Cord blood (CB) has many theoretical advantages that result from the relative immaturity of newborn cells. Compared to adult cells, hematopoietic progenitors from cord blood are enriched in the most primitive stem cells, producing in vivo long-term repopulating stem cells [1,2]. Cord blood hematopoietic stem cells grow larger colonies in presence of growth factors and have different growth factor requirements, with an autocrine production of some growth factors. In presence of growth factors, in vitro long-term culture gives significantly higher yields of hematopoietic progenitors. Cord blood stem cells engraft severe combined immunodeficiency–human (SCID–human) mice in the absence of additional human growth factors. Telomeres are longer than in adult cells [3–5].

The second advantage of cord blood is the relative immaturity of the immune system at birth [6]. Cord blood lymphocytes are enriched in double-negative CD3+ cells and have a naive phenotype. They produce fewer cytokines and express mRNA transcripts for interferon-γ (INF-γ), IL-4, and IL-10 but very little IL-2. They have a fully constituted polyclonal T-cell repertoire and could be protected from apoptosis because of low levels of CD95. Also, the natural killer (NK) function is reduced.

Most of these functions are inducible through in vitro or in vivo activation. As a consequence, early NK and T-cell cytotoxicity is impaired but secondary activation can occur [7]. Acute GVHD is an early event after allogeneic bone marrow transplantation (BMT) and is triggered in part by cytokine release; thus, it is reasonable to postulate that cord blood transplant induces less frequent and less severe acute GVHD than an adult hematopoietic stem cell transplant, which contains a higher number of activated T cells. Therefore, we can speculate that GVHD should be reduced after matched or mismatched cord blood transplant but that graft-versus-leukemia (GVL) should be maintained because of the presence of precursors of T and NK cells [8–10]. These properties should lead to less stringent criteria for HLA matching for donor–recipient selection.

The main practical advantages of using cord blood as an alternative source of stem cells are the relative ease of procurement, the absence of risks for mothers and donors, the reduced likelihood of transmitting infections, particularly cytomegalovirus (CMV), and the ability to store fully tested and HLA-typed blood, in the frozen state, available for immediate use. These advantages were first recognized in cord blood transplant using related donors; secondarily, cord blood banks for unrelated donor transplants established criteria for standardization of cord blood collection, banking, processing, and cryopreservation for delivering units for the treatment of various hematological malignant and nonmalignant diseases [11,12]. More than 25 000 units of frozen cord blood are currently available, and this amount is rapidly increasing worldwide. Since the first cord blood transplant was performed in 1988 [13], cord blood transplantation has been increasingly used as a new source of hematopoietic stem cells, and more than 1 000 cases have been reported worldwide.

To develop and evaluate cord blood transplant results, the European Blood and Marrow Transplantation group (EBMT) has organized a concerted action, the Eurocord group. The objectives of Eurocord are to study the properties of hematopoietic progenitors and gene transfer in cord blood; to study the immune function of cord blood lymphocytes; to coordinate and facilitate the exchange of sera and cells from donor and recipients of cord blood transplants; to establish a European registry of patients treated by cord blood transplants; and to design protocols comparing cord blood transplants with alternative conventional blood and bone marrow hematopoietic stem cell transplants.

Recently, Netcord, a nonprofit organization, was created to establish criteria of qualification and accreditation of cord blood banks and a network was set up between cord blood banks for facilitating donor search. The purpose is to ensure the quality of the cord blood units obtained for transplantation and to aid transplant centers in finding suitable units so as to increase the number of cord blood transplants throughout the world.

Cord Blood Banking

Quality System for Umbilical Cord Blood Banking

It is important to ensure that placental blood banks provide patients with safe and effective products that conform to predefined characteristics. An approach to achieve this target is to implement a quality control system covering all steps of the process involved in the production of umbilical cord blood, from the donor to the patient. In the bank, such a system encompasses the quality of premises, processing, equipment, reagents, staff selection, training, and continuing education. Moreover, for such a system to be successful, it cannot grind to a halt at the placental blood bank door, but it must be continued through to the bedside and involve the clinicians in charge of transplantation.

ISO 9002 guidelines are part of the ISO 9000 series of quality standards established by the International Organization of Standardization. In 1995, the Italian network of cord blood banks, called GRACE (the acronym of Gruppo per la Recollta e Amplificatione delle Cellule Emopoietic), has chosen to operate in compliance with ISO 9002. Members of GRACE include the Turin, Rome, Florence, and Milano cord blood banks. The Milano cord blood bank was ISO 9002 certified in 1997 and acts as the coordinator of GRACE. As Netcord was established as a result of the positive GRACE experience, Netcord members including Dusseldorf, New York, Paris, London, Barcelona, and Milano will also use 9002 standards as a reference in establishing a quality system for cord blood banking. Pr. Girolano Sirchia, managing director of Netcord, has recently published an extensive report on the implementation of a quality system (ISO 9000 series) for placental blood banking [14].

Collection and Cell Processing

Collection can be performed immediately after delivery by a midwife when the placenta is still in utero; cord blood may also be collected from the placenta in an adjacent room by dedicated personnel from the cord blood bank. The advantage of the first method is that the volume of cells collected is higher if the cord is clamped early and the collection begun immediately, but this method can disrupt the normal process of delivery. The second technique is easier, but it might decrease the number of cells collected and perhaps increase the risk of bacterial contamination or clotting. Because of the small volume of cord blood, ranging from 40 to 150 ml, there is some concern that any attempt at cell manipulation and concentration might result in a considerable cell loss of stem cells, which might impair engraftment. Many banks freeze, in a programmed cell freezer, whole cord blood in 20% DMSO; others use HES (hydroxyethyl starch) sedimentation for volume reduction and removal of red cells [11,12]. CD34+ cell selection and stem cell expansion are currently being investigated, mostly for gene transfer experiments or use of cord blood transplants in adults. Thawing technique is well established; it aims at removing red cells and DMSO [11].

Evaluation of stem cell content is a very important issue because several studies have shown that there is a correlation between the number of cells infused and engraftment. There is also a correlation between placental weight, time of clamping, speed of processing, volume collected, and stem cell content. Quantification of hematopoietic stem cells in cord blood is not always easy; most studies refer to nucleated or mononucleated cells infused per kilogram after thawing. Enumeration of CD34+ cells by FACS (fluorescence-activated cell sorter) analysis has entered into routine practice in most laboratories, but results are not always reproducible. Others count the number of CFU-GM in clonogenic assays.

Infectious Diseases Screening

Syphilis and viral tests including HIV-1 and -2, hepatitis B and C, and CMV are performed on mother blood, In some countries, HTLV-1 and toxoplasmosis are also tested. Most of the time, virology tests are not performed directly on cord blood, but a separate aliquot is kept to perform these tests before transplant. Considerable discussion has been engaged on the risk of collecting infected cord blood during the window period before seroconversion. When possible, cord blood is quarantined until the mother has a confirmatory test 3 to 6 months after delivery. In other cases, sensitive tests, including viral antigen or nucleic acid screening tests, when performed on a stored aliquot of cord blood before transplant should considerably decrease the risk of viral transmission [15]. The frequency of bacterial contamination diminishes with the expertise of the staff in charge of the collection. In all cases, bacterial culture for anaerobic and aerobic bacteria must be performed and results sent to the transplant physician when the cord blood is delivered for transplant.

Genetic Diseases

Familial history, ethnic background, and follow-up of the donor should give indications on the risk of transmitting a genetic disease. Tests on cord blood are expensive, and there is no real consensus on the type and number of tests that should be performed. Also, there are some concerns about the notification of the results to the family.

HLA and Red Cell Typing

HLA typing is performed on an aliquot of cord blood. Usually, HLA-A and HLA-B antigens are identified by serology and HLA-DRB1 by DNA amplification methods. Some banks routinely type the mother for HLA to have information on the haplotype and to control the accuracy of cord blood typing.

Ethical and Legal Aspects

It is generally agreed [16–22] that cord and placental blood is a discarded product that can be used without asking permission, but an informed consent must be signed for doing communicable diseases tests on the mother and to the cord blood. The mother must also accept the principle of free and anonymous donation. A recent consensus statement concluded that umbilical cord blood (UCB) is still investigational: during this investigational phase, secure linkage should be maintained of stored UCB to the identity of the donor; UCB banking for autologous use is associated with even greater uncertainty; marketing practices for UCB banking in the private sector need close attention; more data are needed to ensure that recruitment for banking and use of UCB are equitable; and the process of obtaining informed consent for collection of UCB should begin before labor and delivery [22].

Clinical Results

Related Cord Blood Transplantation

Eurocord has analyzed the outcome of 102 patients transplanted with a related cord blood between October 1988 and January 1998 [23–27]. The median follow-up time was 30 months (range, 0.1–111). The median age was 5 years (0.2–20) and the median weight 19 kg (5–50). Sixty-one patients had a malignant disease, including 43 acute leukemia (AL); 25 were in first or second complete remission (CR), and 18 were in post-second CR or resistant relapse; 7 had chronic myeloid leukemia (CML), 7 myelodysplastic syndrome (MDS), 2 neuroblastoma, and 2 non-Hodgkin lymphoma (NHL). Forty-one patients had a nonmalignant disease; 19 patients had aplastic anemia (AA), 15 had hemoglobinopathies, and 7 inborn errors. The donor was an HLA-identical sibling in 80 cases and an HLA-mismatched donor in 22 cases. Five patients had one HLA difference, 6 patients had two HLA differences, 10 patients had three HLA differences, and 1 patient, four HLA differences. The median number of nucleated cells (NC) infused per kilogram was 4.0×10^7 (0.7–18×10^7). Median time to neutrophil engraftment was 28 days (8–49) and 48 days (14–180) for platelets engraftment. Eighteen patients did not engraft, and 10 died before day 60.

The 1-year overall survival was 64% ± 5%. HLA-identical transplants had a 1-year survival of 73% ± 5%; it was 30% ± 10% in patients transplanted with HLA-mismatched cord blood ($p = 0.0006$). According to the initial diagnosis,

1-year survival was 55% ± 7% in patients with malignancies, 67% ± 11% in patients with AA, 100% in patients with hemoglobinopathies (but 49% of disease-free survival, DFS) and 71% ± 17% in patients with inborn errors. Other factors influencing survival favorably were age 6 years or less, weight 20 kg or less, negative recipient cytomegalovirus (CMV) serology, sex match, and number of cells infused of 3.7×10^7/kg or more. There was a relationship between the number of cells infused and engraftment. The incidence of grade II GVHD or greater was 24% ± 4% (7 patients had acute GVHD grade III–IV), and chronic GVHD was observed in 3 of 43 patients at risk. HLA differences were the major prognostic factor for acute GVHD; HLA-identical cord blood transplantation (CBT) had an incidence of acute GVHD of 9% compared to 50% in mismatched CBT ($p = 0.001$).

We can conclude from this study that sibling cord blood transplant gives results comparable to bone marrow transplant, but this must be confirmed by a comparative study. The group of patients receiving an HLA-mismatched CBT is very interesting, as we found a relationship between the number of HLA mismatches and GVHD. However, the sample size should be increased before making any definitive conclusion on the relative importance of HLA disparities.

Unrelated Cord Blood Transplantation

In April 1998, 158 unrelated cord blood transplants were reported to Euro-cord [23,25,27–29], 112 children and 46 adults. One hundred and two children given an unrelated CBT from July 1994 to January 1998 were analyzed. The median age was 5 years (range, 0.2–14), median weight, 19 kg (range, 5–46), ad median follow-up time, 12 months (range, 0.3–42). Seventy-two patients had a malignant disease, including 40 ALL, 20 AML, 5 MDS, 5 juvenile CML, and 2 NHL; 12 patients had AA and 18 patients had an inborn error. Cord blood was provided by an European cord blood bank in 59 cases and by the New York cord blood bank in 43 cases. The median number of NC infused was 4×10^7/kg. Fourteen patients received an HLA-matched CBT; 64 had one HLA difference, 23 had two HLA differences, and 1 had three HLA differences, as defined by serology for class I and high molecular resolution for HLA-DRB1.

The overall 1-year survival was 35% + 6%. Factors associated with improved survival were CMV-negative status before CBT ($p = 0.02$) and ABO match ($p = 0.01$). In patients with malignancies, the overall 2-year survival was 32% ± 6%. The most important favorable factor influencing relapse, DFS, and survival in children with acute leukemia was a good risk disease at transplant (first and second complete remission) [26]. In AA, only 10% of the patients survived [27] and in those with inborn errors survival was 70% ±

11%. Age, weight, number of cells infused, HLA, and sex differences were not significantly associated with survival. Neutrophil engraftment was observed in 74% of the cases. The incidence of acute GVHD grade II or higher was 38%; there was no correlation between the number of HLA mismatches and GVHD.

In 44 adults, including 42 malignancies [29], the overall survival was 16% ± 6%. Most of the patients received fewer than 3×10^7/kg nucleated cells; the donor was HLA identical in only 2 cases, and the others had various degrees of HLA mismatches. The number of cells infused and disease status before transplant were the main prognostic factors, but the heterogeneity of the diagnosis, given the sample size, precludes any statement on the indications and methods of CBT in adults. Of note, the use of steroids was a good prognostic factor while the use of methotrexate (MTX) was associated with poorer survival.

These results show that unrelated cord blood transplant gives good results in children mainly if they are transplanted for acute leukemia in remission, inborn errors, and immune deficiencies. Results must improve in patients with aplastic anemia and in adults. HLA disparity is not a limiting factor, but the number of cells infused is important; currently the use of a number of nucleated cells less than 1×10^7/kg is not recommended. Several questions remain including the criteria of choice of the donor, the indications of CBT in children and in adults, the comparison of cord blood transplants with other sources of hematopoietic stem cells, and the role of growth factors and expansion for improving the speed of engraftment.

Discussion

With the increased number of cord blood transplants being analyzed, it is time to try to answer to the questions that will help making the choice of the source of stem cells in various clinical situations.

Are There Enough Stem Cells in a Single Cord Blood for Short- and Long-Term Engraftment?

The first question about the general use of cord blood for allogeneic hematopoietic stem cell transplantation has been the concern about the engraftment potential of a single cord blood unit in patients with all hematological conditions and all weights. We and others [30–33] have shown that a high number of nucleated cells infused is a good prognostic factor for engraftment and survival. We found that the number of NC infused/kg after thawing was a major factor for predicting recovery of neutrophil and platelet counts after transplantation. In unrelated CBT, patients who received less than

3.7×10^7 NC/kg had a median time to reach 500 µl neutrophils of 34 days (range, 14–48 days) and a median time to reach 20 000/µl platelets of 134 days (range, 30–180 days), while for patients who received more than the median cell dose the median times were, respectively, 25 days (range, 10–56 days) for ANC and 47 days (range, 9–85 days) for platelet recovery. In adults, we found that none of the patients who received less than 1×10^7 NC/kg survived. This result shows that a low number of nucleated cells infused was associated with both an increased risk of nonengraftment and a delay of engraftment. The number of CD34+ cells present at collection and after thawing might also be a good indicator for engraftment [32]. Unfortunately, efforts at standardization of CD34 counts have failed, and therefore could not be included in our analysis.

It is interesting to note that the number of cells infused is 1 log less than a standard allogeneic bone marrow transplantation and 10 times less than a standard peripheral blood stem cell transplantation. In consequence, the number of cells infused is far less than the recommended dose of bone marrow cells; this finding is in favor of the hypothesis that cord blood cells have a selective qualitative advantage on adult bone marrow cells. We have also shown that other factors interfere with engraftment such as diagnosis and HLA differences. Patients with severe aplastic anemia [27] or with hemoglobinopathies [26] had less engraftment than patients with leukemia or inborn errors. This is, as for allogeneic bone marrow transplantation, caused by the addition of several factors including transfusion immunization and reduction of the conditioning myeloablative regimens used for transplantation in nonmalignant disorders. HLA disparities can influence the kinetics and probability of neutrophil and platelet engraftment [23,33].

To improve the speed of engraftment, several methods can be investigated, such as the use of hematopoietic growth factors such as G-CSF, kit ligand, or thrombopoietin (TPO). At this stage, the usefulness of these factors has not been demonstrated and deserves further investigation. Another approach could be to expand in vitro cord blood progenitors to improve short-term engraftment. This area of investigation seems particularly interesting, as in vitro studies have shown that expansion was increased in cord blood compared to bone marrow cells. On the practical point of view, these findings are the basis for the recommendation to blood banks to collect as many cells as possible and to improve the technique of processing and volume reduction to have units available with high cellular counts. The other consequence is that, at this stage, we do not recommend transplanting a patient with units containing, before thawing, fewer than 2×10^7 NC/kg knowing that the estimated loss of cells during thawing has been observed at 20% of the original count before processing. Finally, protocols for studying cord blood stem cell expansion for improving short-term engraftment could be clinically useful.

Is GVHD Reduced After Cord Blood Transplantation?

One of the first concern raised by the use of cord blood for allogeneic transplant was the possibility of inadvertent transplant of cells of maternal origin. We, and other authors, have shown that indeed maternal cells were always present in cord blood but that their number was insufficient to induce GVHD [34,35]. Their presence in high numbers can be detected in the cord blood by HLA high molecular resolution typing, which can show a double population. For these reasons, the detection of maternal cells has not played a role in the practice of cord blood banking. In addition, nobody has shown, so far, engraftment of maternal cells after cord blood transplantation; this suggests that either these cells are in small number or that they are functionally inactive.

Immunological immaturity of CB cells might decrease the incidence and severity of acute GVHD even in the HLA-mismatched situation. At this time, there is no case-control study comparing the incidence of GVHD according to the stem cell source, but there is some evidence that GVHD could be reduced after cord blood transplantation. As in previously published CBT series, we observed GVHD, but it was less severe than in BMT and the incidence of chronic GVHD was low. These findings are quite remarkable for highly mismatched transplants without T-cell depletion. In a cooperative Eurocord-IBMTR study [36], we compared the incidence of acute and chronic GVHD in children receiving either a cord blood or a bone marrow transplant from an HLA-identical sibling donor. In univariate and multivariate analysis, we have shown that both acute and chronic GVHD incidence were significantly reduced after cord blood transplant compared to bone marrow transplant. The incidence of acute GVHD was 24% ± 2% after BMT and 15% ± 7% after CBT (RR, 0.6; $p = 0.04$). The incidence of chronic GVHD was 16% ± 2% after BMT and 6% ± 6% after CBT (RR, 0.3; $p = 0.01$).

The analysis of unrelated cord blood transplants is more difficult because most transplants had several numbers of HLA mismatches and because there were differences of techniques used for typing. We found that HLA incompatibilities were associated with acute GVHD in related transplants. In contrast, we did not find any correlation between the number of mismatches and GVHD in unrelated cord blood transplants, perhaps because we did not have high-resolution typing, except for HLA-DRB1, which makes interpretation of the results difficult.

Is Graft-Versus-Leukemia Diminished After Cord Blood Transplantation?

As cord blood lymphocytes are decreased in number and are immunologically immature, they might lose their GVL function. Experimental data suggest that indeed there is some loss, but on the contrary we have shown that NK cells and T cells in cord blood were inducible by allogeneic stimulation. In patients, in the absence of comparative studies with other sources of stem cells, GVL effect is difficult to assess as the number of patients is limited and the follow-up is too short. An interesting observation was the GVL effect without GVHD of donor lymphocyte infusions after relapse in related CBT recipients.

What Is the Place of Cord Blood Transplant Compared to Other Sources of Hematopoietic Stem Cells?

Current results show that cord blood has proven effective in treating children with malignancies and other disorders. There are not enough patients in other disease categories and in adults to perform a comparison with other stem cell sources. In children with acute leukemia, related and unrelated, and matched or mismatched cord blood transplants gave similar results [25], the main prognostic factor being the stage of the disease at time of transplantation. This cohort of patients compares favorably with other results in the literature [37–39], and a case-control study is currently being performed

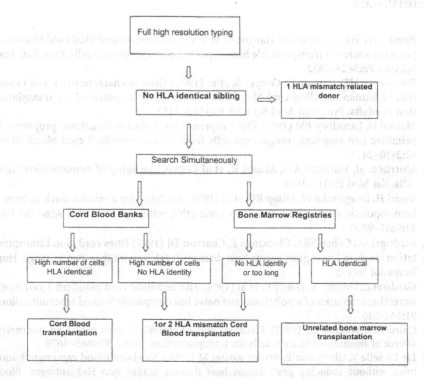

FIG. 1. Strategies for searching alternative hematopoietic stem cell donors

with the European Blood and Marrow Transplant group (EBMT) and selected single-center groups. The Eurocord-IBMTR study [36] comparing survival in children receiving an HLA-identical sibling cord blood or bone marrow transplant does not show any statistical differences between the groups showing that cord blood transplant is as effective as a bone marrow transplant in children.

At this stage and in the absence of definitive answer, we recommend searching simultaneously Bone Marrow Donor Registries and Cord Blood Banks. The final decision must take into account the degree of HLA identity, the availability of the donor, the speed of search, the urgency of the transplant, the number of cells present in the cord blood, and in the case of unrelated bone marrow donor transplants the donor age, sex, number of pregnancies, and CMV status. Prospective studies of the algorithm of donor search will be very important to study the cost efficiency of both procedures (Fig. 1).

Acknowledgment. The authors represent the Eurocord Cord Blood Transplant Group Eurocord is supported by a European concerted action BIOMED II (BMH4-CT96-0833) grant.

References

1. Broxmeyer HE, Gordon GW, Hangoc G, et al (1989) Human umbilical cord blood as a potential source of transplantable hematopoietic stem/progenitor cells. Proc Natl Acad Sci USA 86:3828–3832
2. Broxmeyer HE, Hangoc G, Cooper S, et al (1992) Growth characteristics and expansion of human umbilical cord blood and estimation of its potential for transplantation in adults. Proc Natl Acad Sci USA 89:4109–4113
3. Mayani H, Lansdorp PM (1994) Thy-1 expression is linked to functional properties of primitive hematopoietic progenitors cells from human umbilical cord blood. Blood 83:2410–2417
4. Morrison SJ, Wandycz AM, Akashi K, et al (1996) The aging of hematopoietic stem cells. Nat Med 2:1011–1016
5. Vaziri H, Dragowska W, Allsop RC, et al (1994) Evidence for a mitotic clock in human hematopoietic stem cells: loss of telomeric DNA with age. Proc Natl Acad Sci USA 91:9857–9860
6. Madrigal JA, Cohen SBA, Gluckman E, Charron DJ (1997) Does cord blood transplantation result in lower graft versus host disease? It takes more than two to tango. Hum Immunol 56:1–5
7. Garderet L, Dulphy N, Douay C, et al (1998) The umbilical cord blood αβ T cell repertoire: characteristics of a polyclonal and naive but completely formed repertoire. Blood 91:340–346
8. Cairo MS, Wagner JE (1997) Placental and/or umbilical cord blood an alternative source of haemopoietic stem cells for transplantation. Blood 90:4665–4678
9. De La Selle V, Gluckman E, Bruley-Rosset M (1996) Newborn blood can engraft adult mice without inducing graft versus host disease across non H-2 antigens. Blood 87:3977–3983

10. De La Selle V, Gluckman E, Bruley-Rosset M (1998) Graft versus host disease and graft versus leukemia effect in mice grafted with newborn blood. Blood 92:3968–3975
11. Rubinstein P, Rosenfield RD, Adamson JW, Stevens CE (1993) Stored placental blood for unrelated bone marrow reconstitution. Blood 81:1679–1690
12. Rubinstein P, Dobrila L, Rosenfield RE, et al (1995) Processing and cryopreservation of placental/umbilical cord blood for unrelated bone marrow reconstitution. Proc Natl Acad Sci USA 92:10119–10122
13. Gluckman E, Broxmeyer HE, Auerbach AD, et al (1989) Hematopoietic reconstitution in a patient with Fanconi's anemia by means of umbilical cord blood from an HLA-identical sibling. New Engl J Med 321:1174–1178
14. Sirchia G, Rebulla P, Lecchi L, Mozzi F, Crepaldi R, Parravicini A (1998) Implementation of a quality system (ISO 9000 series) for placental blood banking. J Hematother 7:19–35
15. Schreiber GB, Busch MP, Kleinman SH, Korelitz JJ (1996) The risk of transfusion-transmitted viral infections. New Engl J Med 334:1685–1690
16. Gluckman E (1995) Advantages of using fetal and neonatal cells for treatment of hematological diseases in human. In: Gluckman E, Coulombel L, (eds) Ontogeny of hematopoiesis: aplastic anemia. Colloque INSERM, vol 235. Libbey Eurotext, Paris, pp 183–190
17. Gluckman E, O'Reilly RJ, Wagner J, Rubinstein P (1996) Patents versus transplants. Nature (Lond) 382:108
18. Sugarman J, Reisner EG, Kurtzberg J (1995) Ethical aspects of banking placental blood for transplantation. JAMA 275:1783–1785
19. Sugarman J, Kaalund V, Kodish E, Marshall MF, Reisner EG, Wilfond BS, Wolpe PR (and the working group on ethical issues in umbilical cord blood banking) (1997) Ethical issues in umbilical cord blood banking. JAMA 278:938–943
20. Marshall E (1996) Private cord blood banks raise concern. Science 271:587
21. Brenner M (1996) Placental blood transplant: who will benefit? Nat Med 2:969–970
22. Gluckman E (1996) The therapeutic potential of fetal and neonatal hematopoietic stem cells. New Engl J Med 335:1839–1840
23. Gluckman E, Rocha V, Boyer Chammard A, et al (1997) Outcome of cord blood transplantation from related and unrelated donors. New Engl J Med 337:373–381
24. Rocha V, Chastang CL, Souillet G, et al (for the Eurocord transplant group) (1998) Related cord blood transplants: the Eurocord experience of 78 transplants. Bone Marrow Transplant 21(suppl 3):S59–S65
25. Locatelli F, Rocha V, Chastang C, et al (1998) Cord blood transplantation for children with acute leukemia. Bone Marrow Transplant 21(suppl 3):S66–S70
26. Miniero R, Rocha V, Saracco P, et al (on behalf of Eurocord) (1998) Cord blood transplantation in hemoglobinopathies. Bone Marrow Transplant 2(suppl 1):S78–S79
27. Rocha V, Chastang CL, Pasquini R, Nagler A, Garnier F, Saarinen U, Arcese W, Boiron JM, Stary J, Veyes P, Fernandez M, Milovic V, Akinoshita K, Favre C, Kremens B, Marin G, Gluckman E (1998) Cord blood transplant in bone marrow failure syndromes (abstract 547). Blood 92(10, suppl 1):136a
28. Gluckman E, Rocha V, Chastang C (1998) European results of unrelated cord blood transplants. Bone Marrow Transplant 21(suppl 3):S87–S91
29. Rocha V, Chastang CL, Laporte JP, Garnier F, Fernandez M, Oteyza JP, Abecasis M, Gluckman E (1998) Unrelated cord blood transplant in adults with malignancies (abstract 580). Blood 92(10, suppl 1):144a
30. Kurtzberg J, Laughlin M, Graham L, et al (1996) Placental blood as a source of hematopoietic stem cells for transplantation into unrelated recipients. New Engl J Med 335:157–166

31. Wagner JE, Rosenthal J, Sweetman R, et al (1996) Successful transplantation of HLA matched and HLA mismatched umbilical cord blood from unrelated donors: analysis of engraftment and acute graft versus host disease. Blood 88:795–802

32. Wagner JE, DeFor T, Rubinstein P, Kurtzberg J (1997) Transplantation of unrelated donor umbilical cord blood: outcomes and analysis of risk factors (abstract 1767). Blood 90(suppl 1):398a

33. Rubinstein P, Carrier C, Scaradavou A, et al (1998) Outcomes among 562 recipients of placental-blood transplants from unrelated donors. New Engl J Med 339:1565–1577

34. Petit T, Dommergues M, Socié G, et al (1997) Detection of maternal cells in human fetal blood during the third trimester of pregnancy using allele specific PCR amplification. Br J Haematol 98:767–771

35. Socié G, Gluckman E, Carosella E, Brossard Y, Lafon C, Brison O (1994) Search for maternal cells in human cord blood by polymerase chain reaction amplification of two minisatellite sequences. Blood 83:340–344

36. Rocha V, Wagner J, Sobosinski K, Horowitz M, Guckman E (1998) Comparative study of graft versus host disease in HLA identical sibling cord blood or bone marrow transplant in children (abstract 2822). Blood 92(10, suppl 1):685a

37. Hongeng S, Krance RA, Bowman LC, et al (1997) Outcomes of transplantation with matched-sibling and unrelated donor bone marrow in children with leukemia. Lancet 350:767–771

38. Szydlo R, Goldman JM, Klein JP, et al (1997) Results of allogeneic bone marrow transplants for leukemia using donors other than HLA identical siblings. J Clin Oncol 15:1767–1777

39. Oakhill A, Pamphilon DH, Potter MN, et al (1996) Unrelated donor bone marrow transplantation for children with relapsed acute lymphoblastic leukaemia in second complete remission. Br J Haematol 94:574–578

Key Word Index